EXTENDING SCIENCE

13

SPORT

Selected Topics

R. B. Arnold B Tech

Stanley Thornes (Publishers) Ltd

First published in 1988 by
Stanley Thornes (Publishers) Ltd
Old Station Drive
Leckhampton
CHELTENHAM GL53 0DN
England

*An exception is made for the puzzles on pp. 38 and 55–6. Teachers may photocopy a puzzle to save time for a pupil who would otherwise need to copy from his/her copy of the book. Teachers wishing to make multiple copies of a word puzzle for distribution to a class without individual copies of the book must apply to the publishers in the normal way.

British Library Cataloguing in Publication Data

Arnold, R.B. (R. Brian)
    Extending science: sport.
    1. Sports & games. Physiological aspects
    I. Title
    612′.044

    ISBN 0–85950–671–1

Typeset by Tech-Set, Gateshead, Tyne & Wear.
Printed and bound in Great Britain by Ebenezer Baylis & Son, Worcester.

CONTENTS

PREFACE

Science can be one of the most exciting subjects pupils study at school. This, however, is only true if pupils perceive it to be relevant and interesting. Sport provides us with many opportunities to see 'science in action'. Understanding how and why things happen in sport will, I am convinced, lead to an even greater enjoyment of sport for both those who like to take part and those who like to watch.

This book is intended to serve three purposes:
1. To help pupils understand some of the basic scientific principles involved in sport.
2. To help and encourage pupils to delve more deeply into their sport(s) and thereby increase their skill level and proficiency.
3. Most importantly I hope pupils will find the book interesting.

R. B. Arnold

ACKNOWLEDGEMENTS

The author and publishers are grateful to the following who provided photographs and gave permission for reproduction:

Colorsport (pp. 1 top right, 5, 8, 11, 44, 75, 77, 82); Dunlop Slazenger (pp. 1 top left, 6 bottom, 30, 47); Ford of Britain (p. 46); James Gilbert (p. 49); Martin Hyndman (p. 37); David Muscroft Photography (p. 62); Sporting Pictures (UK) Ltd (pp. 1 bottom, 76 top); Tony Stone Ltd (p. 6 top); Sally Anne Thompson, Animal Photography Partnership (p. 76 bottom)

FORCES

FORCES IN SPORT

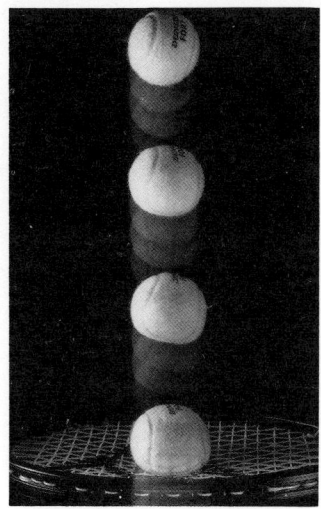

Both ball and racket deform when a shot is played in tennis.

In a sport in which just hundredths of a second can make the difference between winning and coming second a good 'push start' is essential

A powerful header can be used to pass the ball or score a goal

Three effects of forces in sport

Look carefully at the photographs on p. 1. They illustrate very clearly the importance of forces in sport and the three main effects they can have.

- Forces can change the shape of an object.
- Forces can change the speed of an object.
- Forces can change the direction in which an object is moving.

A thorough understanding of forces and how they can be used could help athletes achieve greater proficiency in their sport.

When and where might one of the effects of a force be seen in the following sports?

Basketball
Cricket
Rugby
High jump
Motor car racing
Squash
Boxing
Archery

WEIGHT

One of the most common forces athletes have to deal with is *weight.*

Sumo wrestlers

These Sumo wrestlers are trying to throw each other out of the ring. The heavier a wrestler is the greater the force his opponent needs to apply to achieve this.

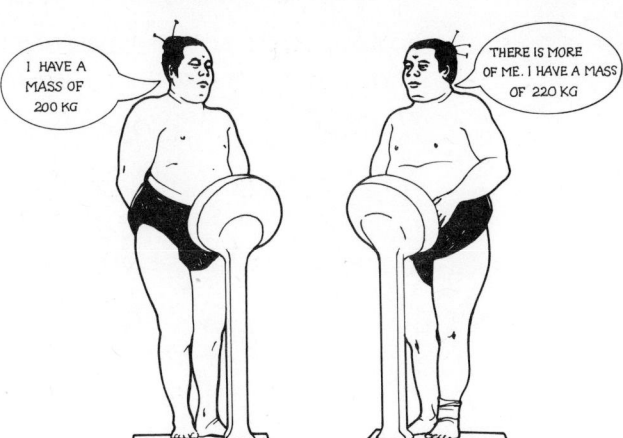

These wrestlers are massive

The two wrestlers above are 'massive' — there is a lot of them! We measure their *mass* in kilograms. In science, we measure weight (and all other forces) in *newtons*. The weight F of an object or person of mass m can be calculated using the equation

$$F = m \times g$$

where F = force
m = mass
g = the acceleration due to gravity (9.81 m/s^2)

How large a force would you feel if you were the one underneath?

If the lighter of the above two wrestlers stood on top of you, how large a force would you feel?

$$F = m \times g$$
$$= 200 \times 9.81$$
$$= 1962 \text{ N}$$

The force you would feel is 1962 N.

Below are given the upper limits for the various weight classes of boxer. Change these values so that they are expressed in newtons.

Flyweight	*50.80 kg*
Bantamweight	*53.52 kg*
Featherweight	*57.15 kg*
Lightweight	*61.24 kg*
Welterweight	*66.68 kg*
Middleweight	*72.58 kg*
Light heavyweight	*79.38 kg*
Heavyweight	*no limit*

FRICTION

Sometimes friction is necessary

Another very common force we need to be aware of in sport is *friction*. Sometimes its presence is necessary for an athlete to perform well. Sometimes, however, its presence is a disadvantage.

Friction may also be a disadvantage

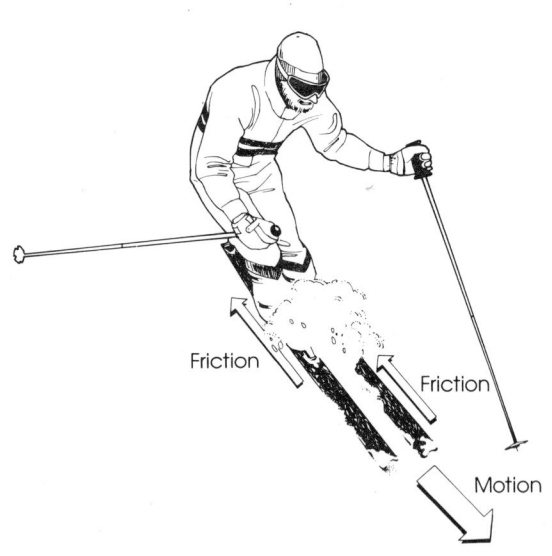

Whenever an object moves or tries to move, friction is present. It always acts in the direction which opposes the motion.

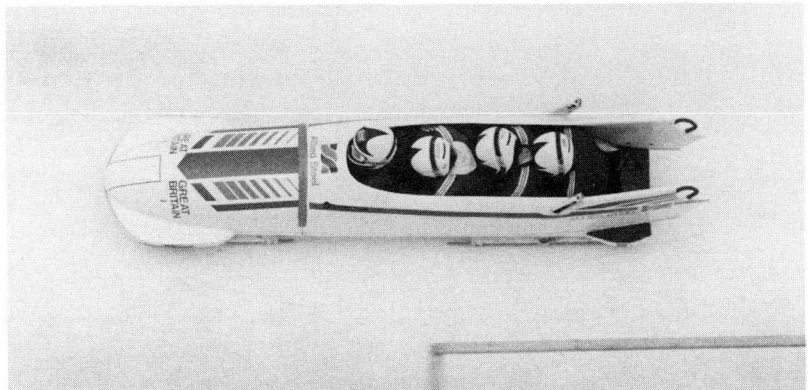

Friction opposes the motion of this bobsleigh

As a bobsleigh travels down a run it gains speed. Frictional forces between the runners and the ice try to prevent this from happening. In order to keep these forces to a minimum the runners are coated with a lubricant such as wax.

The most crucial member of a bobsleigh team is the driver. He must steer the bobsleigh down the narrow run without catching the icy walls. Each time he touches the wall, friction forces slow the team down so that they lose precious hundredths of a second.

Why do bobsleighs have the shape they do?

On two occasions during the run it is essential to have and make use of frictional forces. When and where do these occur?

Whether it is on a squash court or football field or at the beginning of a race it is important for athletes not to lose their footing. These are occasions when friction is useful.

To reduce the possibility of slipping, manufacturers of sporting footwear produce shoes and boots adapted to the needs of individual sports.

Sporting footwear

Look carefully at the shoes/boots shown above. Which sports are likely to use these? Explain your answers.

Modern grand prix racing cars can travel at over 350 km/h or the straight, and will corner at speeds in excess of 200 km/h. One of the reasons they can do this without skidding off the track is that they have special tyres, called *slicks*. These tyres, unlike those on a normal car, have no tread. They use the simple idea that the more of the tyre there is in contact with the track the greater the frictional force (grip) it can make use of as it corners, brakes or accelerates.

Slicks on a modern racing car

If, however, there is any lubricant on the track, e.g. water or oil, this will come between the tyre and the track, drastically reducing the frictional forces between them. As a result the car will almost certainly lose its grip on the road and skid. To overcome this problem special wet-weather tyres are used if the track is at all slippery. These tyres squeeze the water into grooves so that good contact can be made between the tyre and track surface.

Treads on tyres help cope with wet surfaces

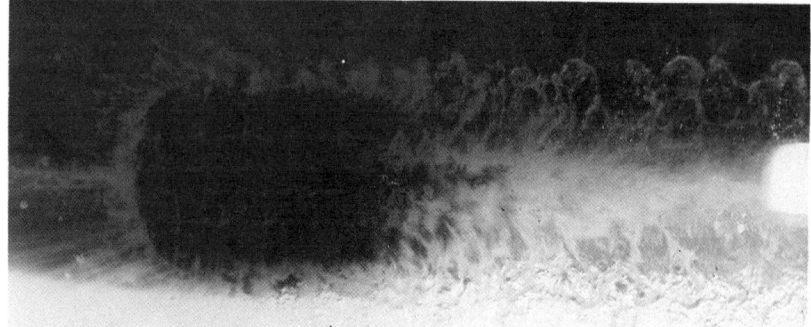

Why don't racing teams use wet-weather tyres all the time?

MUSCLES

We can apply and direct forces because we have muscles in our bodies.

Muscles consist of bundles of fibres. When the brain sends a message to a muscle a chemical reaction takes place which causes the muscle to contract or relax. Muscles which control the movement of limbs are connected to the bones via tendons and arranged in *antagonistic pairs*. In each pair, the muscle that straightens a limb is called the *extensor*, and the one that causes bending or hinging at the joints is called the *flexor*.

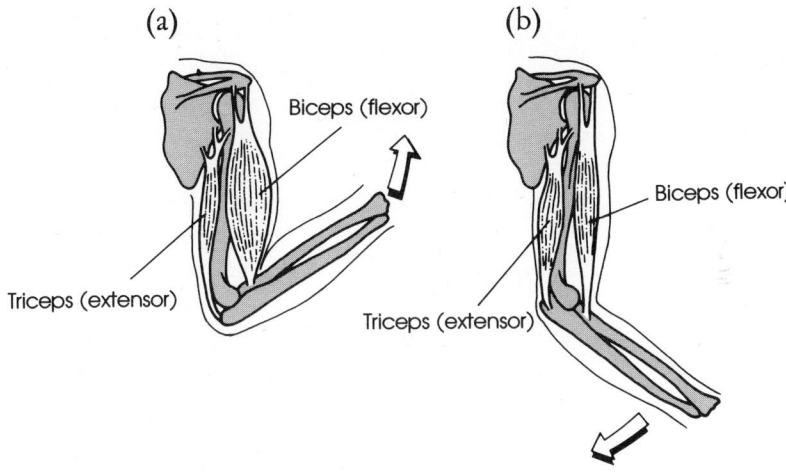

(a) (b)

Biceps (flexor)

Biceps (flexor)

Triceps (extensor)

Triceps (extensor)

Muscles in the arm:
(a) arm bent
(b) arm straight

When the arm is being bent upwards the *biceps* muscle is contracting, whilst the *triceps* muscle is relaxing. When the arm is being straightened the biceps relaxes and the triceps contracts. This co-operative behaviour of muscles is the basis of movement.

ACTIVITY 1

3
4
1
2
Paper fastener

1) Draw these two shapes on a piece of stiff card.

2) Cut the shape out and make four small holes in the positions 1, 2, 3 and 4.

7

3) Using a paper-fastener, fix the two halves of the arm together.

4) With the pieces of the arm positioned as shown, attach two pieces of elasticated thread so that both are slightly stretched.

5) Straighten the arm. What happens to threads A and B?

6) Bend the arm upwards. What now happens to threads A and B?

Model of arm

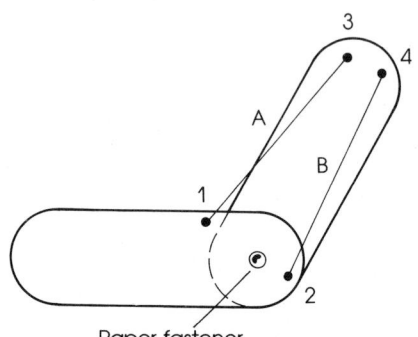

Paper fastener

WORK

Strongman pulling a lorry

The photograph above is of a competitor in one of the events to find the 'World's Strongest Man'. In this event competitors must pull a lorry a distance of 20 metres as quickly as possible. In order to get the lorry moving these strongmen have to exert a force at least equal to the frictional forces trying to prevent it from moving. At the end of the event some of the competitors are so exhausted that they need oxygen. They have been made to do a lot of *work*.

We can calculate how much work each competitor had to do by using the equation

Work done = Force × Distance moved

If the force exerted by each competitor is 2000 N and the distance the lorry is pulled is 20 m,

Work done = 2000 × 20
= 40 000 joules (J)

Calculate how much work has been done in each of these events:

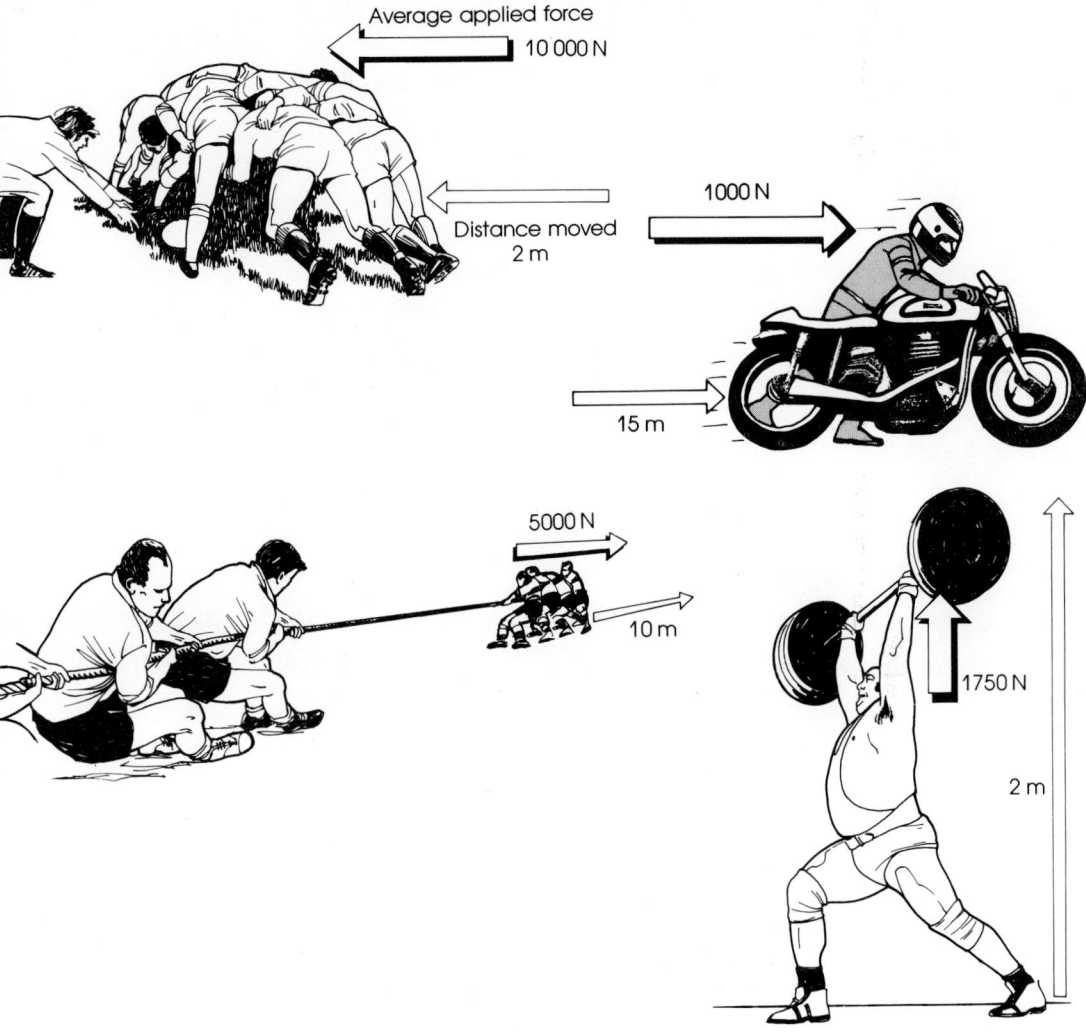

Average applied force
10 000 N

Distance moved
2 m

1000 N

15 m

5000 N

10 m

1750 N

2 m

POWER

To win the event shown on page 8 the strongmen had not only to pull the lorry 20 metres but also to complete the task as quickly as possible. The winner was the man who could do the 40 000 J of work the most quickly. When we are concerned with how rapidly work is being done we are considering *power*.

$$\text{Power} = \frac{\text{Work done}}{\text{Time taken}}$$

If the most powerful competitor completed the event in 25 seconds what is his power rating?

$$\text{Power} = \frac{\text{Work done}}{\text{Time taken}}$$

$$\text{Power} = \frac{40\ 000}{25}$$

$$= 1600 \text{ joules per second (J/s)}$$

$$= 1600 \text{ watts (W)}$$

(1 watt = 1 joule per second)

Which is the most powerful?

Which of these three cars is the most powerful? Explain precisely what this means. Use the word 'work' in your explanation. What forces do the engines of these cars have to pull against?

ACTIVITY 2

How powerful are you?

1) Weigh yourself (in newtons).

2) Using a stable stool or step, step up (both feet on top) and then step down, as many times as you can in 30 seconds.

3) Measure the height of the step (in metres).

4) Use the equation below to calculate your power.

Work done with each step up
= Force (weight) × Height of step

Total work done = Number of steps up × Force × Height

$$\text{Power} = \frac{\text{Total work done}}{\text{Time taken}}$$

Measuring your power

CENTRE OF MASS

High-jumper taking off

The high-jumper shown above has to do work in order to lift his body over the bar. To calculate how much work he must do we can use the equation

Work done = Force × Distance moved

The average force he will need to apply will be approximately equal to his weight in newtons.

The distance moved will be the height through which he has had to lift his weight. The jumper's weight is obviously spread throughout his body and therefore it is difficult to say how large a distance the weight has had to move. To solve this problem we imagine that all the weight of the jumper is concentrated in one place called the *centre of mass* (sometimes also called *the centre of gravity*). Its exact position varies from person to person, depending on build, and also shifts as a person moves. The activity below shows how you can discover the position of your centre of mass.

ACTIVITY 3

1) Weigh yourself.

2) Lie down on a plank of wood. Mark the position of your head and feet.

3) Ask someone to lift one end of the plank off the floor using a newtonmeter. He or she should note down the force needed to do this.

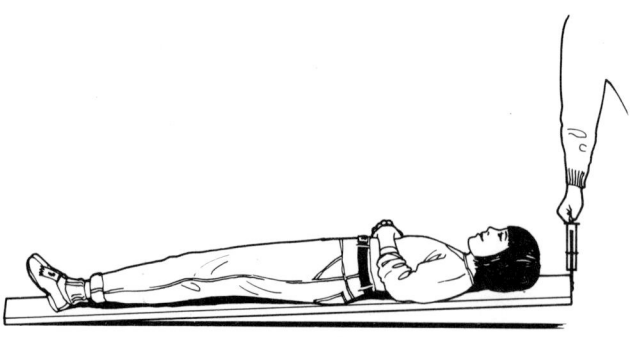

4) Now place weights equal to your weight on the plank.

5) Lift the plank slightly so that one end is off the ground.

6) Move the weights back and forth until the newtonmeter reading is the same as it was when you lay on the plank.

7) Measure the distance from the centre of the weights to the mark indicating where your head was when you lay on the plank. This measurement indicates how far your centre of mass is below your head.

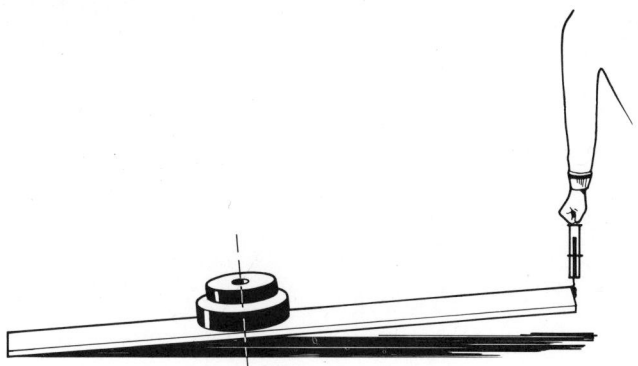

Now calculate the minimum amount of work that needs to be done by this jumper to clear the bar.

Work done = Force × Distance moved

$$= \ldots$$

How much work does the high-jumper do in jumping?

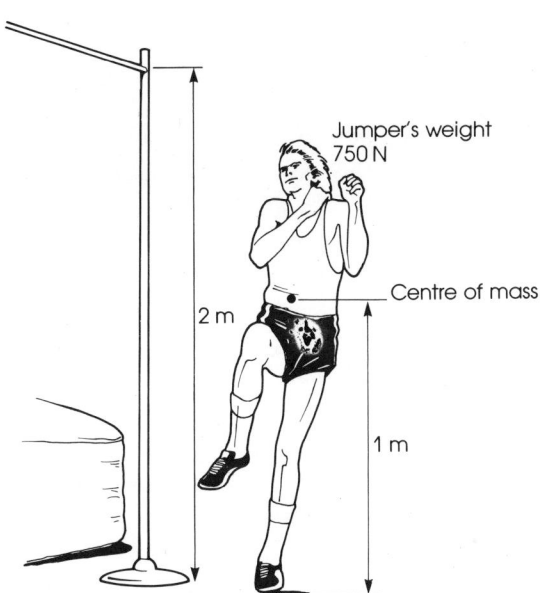

Jumper's weight
750 N

Centre of mass

2 m

1 m

From the above we can see that

- If a jumper is tall his centre of mass will be high and he will therefore clear the bar more easily than a smaller jumper of the same mass.

● The best jumping technique will be the one which allows the jumper to clear the bar without having to raise his centre of mass too high.

Various styles of high-jumping

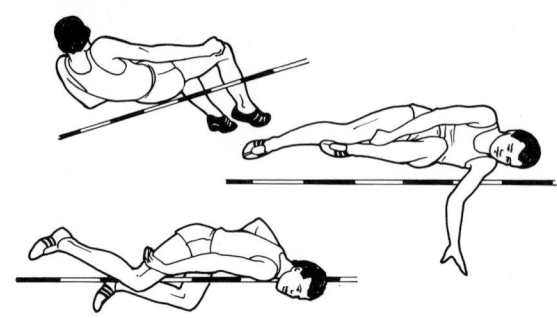

During a race a long-distance runner must try to conserve energy. A running style which causes him to bob up and down will waste energy, as work must be done in raising his centre of mass.

Good running style

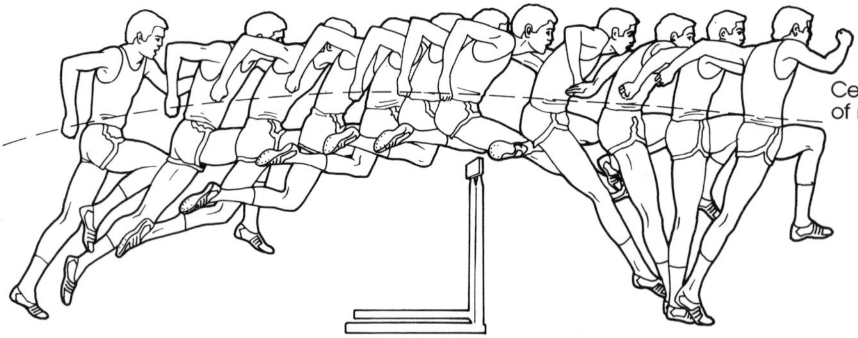

Centre of mass kept level

Hurdling

Centre of mass

Why is this a good hurdling technique? (Hint: *what is happening to the hurdler's centre of mass?*)

ENERGY

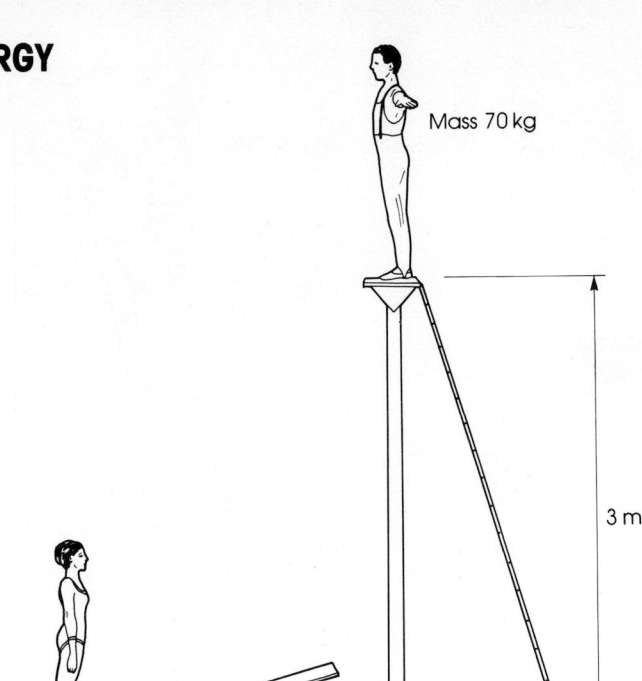

Mass 70 kg

3 m

This acrobat has potential energy

When the above acrobat jumps down on to the see-saw he propels (lifts) his partner upwards. In doing so he is 'doing work'. Someone or something which is capable of 'doing work' is said to possess *energy*. The acrobat has energy because of his position. He would be unable to lift his partner if he was also standing on the ground. Energy possessed by an object because of its position is called *potential energy*.

Energy, like work, is measured in joules. We can calculate the acrobat's potential energy (P.E.) using the formula

$$\text{P.E.} = m \times g \times h$$

where m = mass
g = acceleration due to gravity (9.81 m/s^2)
h = height above ground

$$\text{P.E.} = 70 \times 9.81 \times 3$$
$$= 2060 \text{ J}$$

Calculate the potential energy of the following, assuming that the centre of mass is at the height stated:

- *A 0.8 kg javelin at a height of 20 m above the ground*
- *A 7 kg hammer at a height of 25 m above ground*
- *A 60 kg high jumper clearing the bar set at 2 m*
- *A 75 kg pole-vaulter clearing the bar set at 6 m*
- *A 2 kg discus at a height of 23 m above the ground*
- *A 50 kg trampolinist at a height of 3 m above the ground*

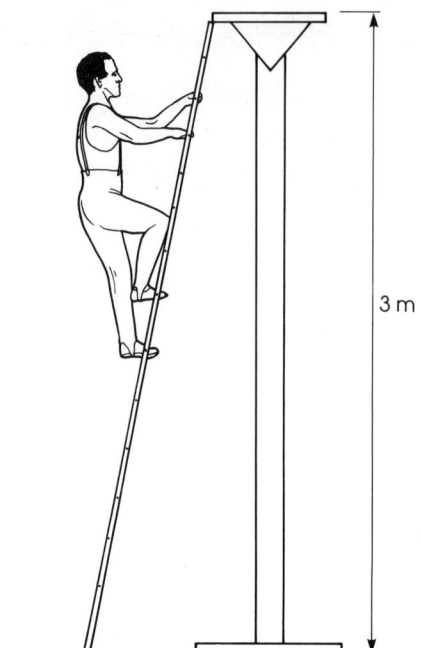

Climbing increases
potential energy

3 m

The acrobat on page 15, in order to possess potential energy, had to climb to the top of his stand. In doing so he did some work. The amount of work he did was equal to the potential energy he gained.

When the acrobat is about to land on the see-saw he has lost nearly all his potential energy. It has been converted into *kinetic energy*. Kinetic energy is energy due to motion.

We can calculate the kinetic energy (K.E.) of an object using the formula

$$\text{K.E.} = \tfrac{1}{2} \, mv^2$$

where m = mass
v = velocity

If the acrobat has a mass of 70 kg and as he lands on the see-saw he is falling at a velocity of 7.7 m/s his kinetic energy is

$$\text{K.E.} = \tfrac{1}{2} \times 70 \times 7.7^2$$
$$= 2060 \text{ J}$$

Calculate the kinetic energy of the following:

- *A 60 kg sprinter travelling at 10 m/s*
- *A 0.8 kg javelin flying through the air with a velocity of 20 m/s*
- *A 0.1 kg arrow flying through the air with a velocity of 80 m/s*
- *A 400 kg bobsleigh travelling at 50 m/s*

CONSERVATION OF ENERGY

As we have already seen with the falling acrobat on page 16 his potential energy does not just disappear, but instead is turned into another form. This situation is a good example of *the law of conservation of energy*, which states that

Energy is neither created nor destroyed but can be converted from one form into another.

Potential energy

Kinetic energy

Pole-vaulting

Pole-vaulting is another event which clearly illustrates this law. If a vaulter sprints down the runway, just before planting the pole he will possess a lot of kinetic energy. Using the pole, he can convert this energy into potential energy, so lifting himself up and over the bar. In an ideal situation his kinetic energy at the end of his run up ($\frac{1}{2} mv^2$) should be equal to his potential energy (mgh) as he clears the bar, i.e.

$$\tfrac{1}{2} mv^2 = mgh$$

We can see from this that sprinting speed as well as good vaulting technique is essential if a pole-vaulter is to do well in competitions.

If a good 100 m sprinter takes up pole-vaulting what height is he theoretically capable of clearing?

If the velocity of the sprinter is approximately 10 m/s then, using the above equation,

$$\tfrac{1}{2} \times m \times 100 = m \times 9.81 \times h$$
$$h = \frac{50}{9.81}$$
$$h = 5.1 \text{ m}$$

This value is actually less than the height being jumped by top pole-vaulters. How is this possible? (*Hint*: how and when can the vaulter put extra energy into his vault?)

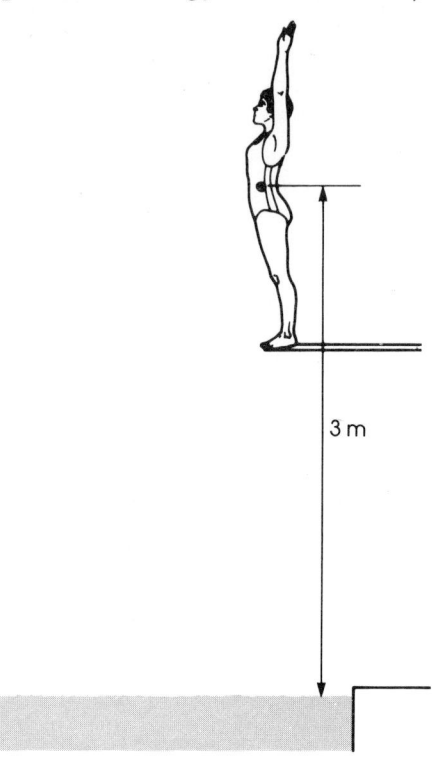

Diver

3 m

This diver has a mass of 60 kg. How much kinetic energy does she have just before she enters the water after diving from the 10 m board? How much kinetic energy does she have half-way down? How much potential energy does she have half-way down?

During a diving competition divers give themselves a special kind of kinetic energy called *rotational* energy. In some sports an understanding of rotational energy and how it can be changed and used is vital. If a diver chooses to do a $2\tfrac{1}{2}$-somersault dive, he or she must control the rate of rotation. To spin faster divers curl up; to slow down they uncurl.

To understand why this happens, let us consider a skater rotating on the spot. If we assume that there is no friction between the ice and the skater, then in accordance with the law of Conservation of Energy the rotational energy of a skater will remain constant. We can calculate the rotational energy of an object using the equation

What then will happen if the skater alters her shape, for example by moving her arms whilst she is spinning?

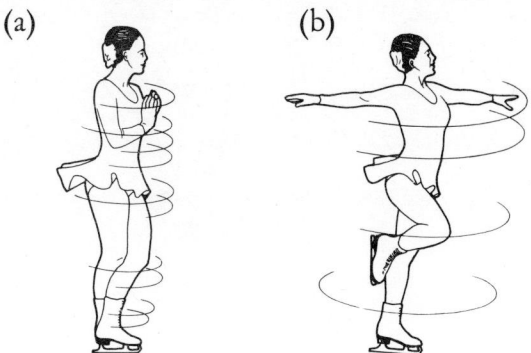
(a)   (b)

Skater spinning:
(a) arms held in
(b) arms held out

If she holds her arms close to her chest her moment of inertia decreases. If her rotational energy is to remain unaltered this must be compensated for by ω becoming larger, i.e. she will spin faster.

Explain what will happen if the skater throws her arms out whilst spinning.

TURNING EFFECT OF FORCES

In Indian wrestling the winner is the one who can push his opponent's hand down on to the table without lifting or moving his elbow. The force exerted by the winner will cause both arms to pivot or turn about the elbows.

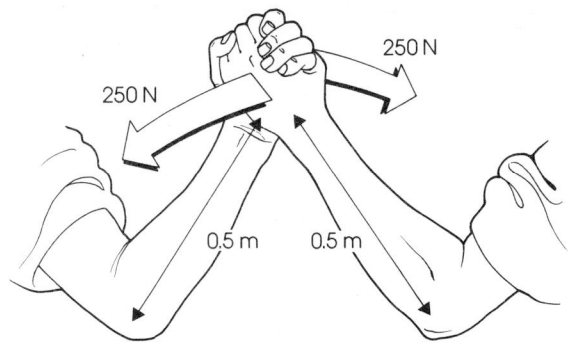

250 N   250 N

0.5 m   0.5 m

Indian wrestling

The turning effect of a force is called a *moment* and its size can be calculated using the formula

Moment of a force = Force × Perpendicular distance from pivot

The moment each wrestler has to resist is approximately

250 N × 0.5 m = 125 N m

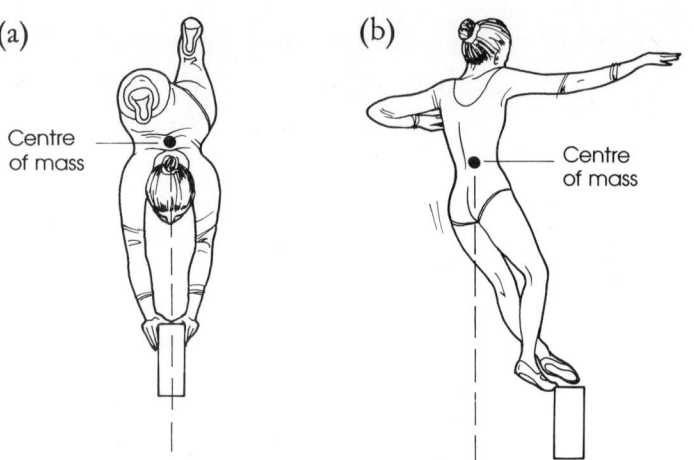

Centre
of mass

Centre
of mass

Gymnast with centre of
mass (a) directly above
beam, (b) not directly
above beam

The gymnast above has a similar but much more subtle problem. When she is positioned such that her centre of mass is directly above the beam she feels balanced and in control. If, however, her centre of mass moves so that it is not directly above the beam her weight creates a moment. If she is unable to resist this, it could cause her to topple from the beam.

How could the gymnast shown above regain her balance if she felt herself begin to topple?

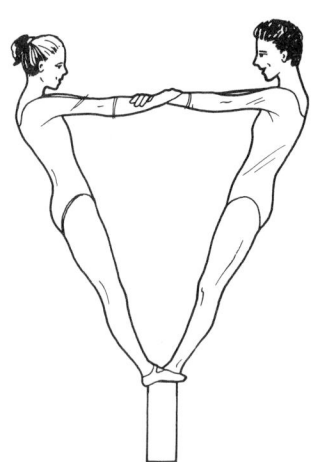

Gymnasts with
moments balanced

Neither of the two gymnasts shown here has her centre of mass directly above the beam, and yet both are balanced. This is because the moment created by the weight of one of them is balanced by an equal and opposite moment created by the other.

If the turning effects of two or more forces are balanced, i.e. there is a state of equilibrium, we can say that

The clockwise moments = The anticlockwise moments

What is wrong here?

What is likely to happen if a weightlifter grips the bar as shown here? Explain your answer.

How should he grip the bar? Explain your answer.

What would happen to this motorcycle and side-car combination if it went round a corner too fast (but without skidding)? What does the passenger in the side-car do to prevent this from happening? What is he creating and why?

Motorcycle and side-car

Explain how moments are created

- When pedalling a bicycle
- When wheeling a scrum
- When doing press ups
- When steering a car
- When using the biceps and triceps muscles in the arm

STABILITY

'Handing off' in rugby

In rugby we often see players trying to 'hand off' one of the opposition. As they do so their push creates a moment which tries to turn or topple the defender. If the defender had his feet close together, and he stood rigidly upright, the turning effect of the 'hand off' would be immediately obvious. In practice, however, this rarely happens. Such a stance would be too unstable, i.e. only a small force would be needed to push the defender over.

Instinctively, would-be tacklers have their feet at least shoulder-width apart and are slightly crouched. This stance is much more stable. The diagrams below show very clearly that a wide base and a low centre of gravity are essential where stability is important.

(a)   (b)

Would-be tackler in
(a) unstable
(b) stable stance

QUESTIONS ON CHAPTER 1

1 What is friction? Give three examples in sport in which friction is an advantage, and three examples in which it is a disadvantage. How can you increase the friction between two objects? How can you decrease the friction between two objects?

2 Look carefully at this simplified diagram showing the muscles in the leg. What happens to the muscles when the leg is straightened? What happens to the muscles when the leg is bent?

What is:

(a) An extensor muscle?

(b) A flexor muscle?

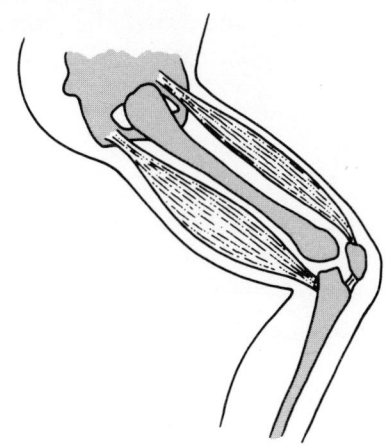

3 This 50 kg trampolinist is at the top of her bounce. What is her potential energy?

What is her kinetic energy:

(a) at this point?

(b) as she comes into contact with the trampoline again?

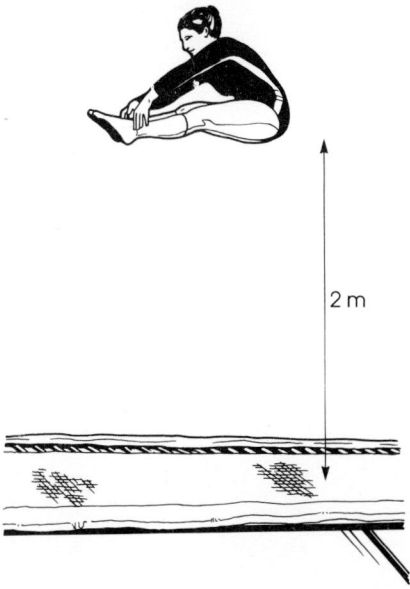

2 m

4 Calculate the work done in the following situations:

(a) A weightlifter lifts a 2000 N weight through a height of 2 m.

(b) Three members of a bobsleigh team push their sleigh 20 m, each of them applying a force of 900 N to it.

(c) A springboard propels a 70 kg diver 1.5 m vertically upwards. ($g = 9.81$ m/s)

(d) Calculate the power (rate of doing work) for (a), (b) and (c) if the time for (a) is 1.5 s, (b) is 4 s and (c) is 0.5 s.

5

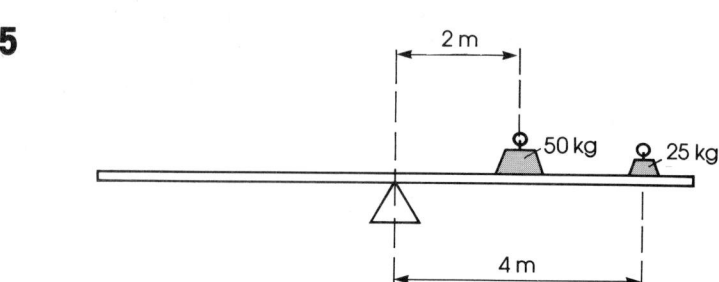

Calculate the total clockwise moment in the above diagram. Where must a 100 kg man stand if he is to balance the beam?

2

FORCES AND
MOTION

INERTIA

Sprinters after a race

When athletes compete in indoor competitions there is less room for the events than there would be outdoors. As a consequence 60 m sprinters often experience great difficulty in stopping quickly once they have crossed the finishing line.

Likewise the skier here, having lost control, was unable to stop himself and ran into the safety fencing and straw bales.

This skier was unable
to stop soon enough

The rugby player on the left below has the opposite problem. One of the opposing team has managed to break a tackle and is sprinting for the line. The defender knows he is a faster runner than the man with the ball but it will take him several seconds to 'get going', i.e. he cannot immediately reach his top speed.

The faster a hammer thrower can move the ball just before release the farther it will travel. The speed of the ball, however, must be built up slowly by 'whirling it around' several times whilst gradually accelerating it. The athlete does this because he is aware that it is impossible to increase the speed of the ball instantaneously.

From the above we can see that objects cannot be accelerated or decelerated instantaneously. They seem 'reluctant' to change their velocity. We describe this property as *inertia*.

The amount of inertia an object possesses depends upon its mass. The greater the mass, the greater the reluctance to change velocity, i.e. the greater the inertia.

ACTIVITY 4

1) Place a 2p coin on a small piece of stiff card on top of a small beaker.

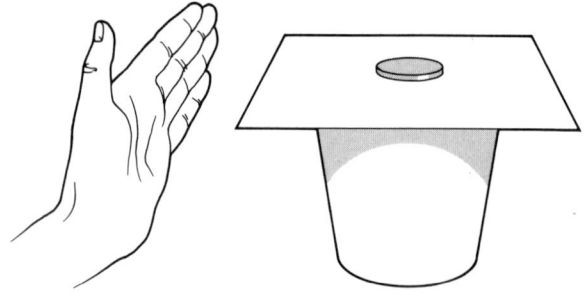

2) Give the card a sharp sideways tap with your finger. Explain what happens. Use the words mass and inertia in your explanation.

3) Repeat the activity with a thin cardboard disc in place of the 2p coin. Explain what happens. Again use the words mass and inertia in your explanation.

In drag car racing competitors must cover the $\frac{1}{4}$-mile course as quickly as possible from a 'standing start'. To overcome their large inertia, these cars often have very powerful rocket engines. What would you expect to happen to the driver of the car when it begins to accelerate?

Like the athlete at the beginning of this chapter competitors in this sport also have to deal with the problem of how to slow down and stop after crossing the finishing line. One solution to this problem is the 'drag chute'.

Drag racing car being decelerated by drag chute

Explain how this works. What would you expect to happen to the driver when the car begins to decelerate?

MOMENTUM

Rugby forwards

The two rugby players seen here are both forwards. Their task is to push the opposition out of the way. At first, we might expect the larger of the two forwards to be the better

27

at this, but he need not be. A player's ability to push an opponent out of the way with his shoulder will depend not only on the mass of the player (m) but also on the velocity with which he is moving (v). If we multiply these two quantities together we obtain the property possessed by all moving objects called *momentum*:

Momentum $= m \times v$

If the larger forward runs with a velocity of 6 m/s, at what speed must the smaller forward run if he is to have the same momentum?

Calculate the momentum of the following:

1)   a) *A 7 kg shot moving with a velocity of 5 m/s.*
     b) *A 0.8 kg javelin moving with a velocity of 15 m/s.*
     c) *A 75 kg sprinter moving with a velocity of 10 m/s.*

2)   *A racing car travelling at 50 m/s has a momentum of 20 000 kg m/s. What is the mass of the car?*

CHANGING MOMENTUM

If we want to change the momentum of an object we need to apply a force. The size of the force will depend upon:

- The size of the change in momentum.
- How quickly the change in momentum takes place.

If the change of momentum is small then only a small force needs to be applied

But . . .

If the change of momentum occurs slowly then only a small force needs to be applied

But . . .

A famous scientist called Isaac Newton suggested that the size of a force could be found using the equation

$$\text{Force} = \frac{\text{Change in momentum}}{\text{Time taken}}$$

$$\text{or } F = \frac{mv - mu}{t}$$

where m = mass of object
u = initial velocity of object
v = final velocity of object
t = time during which the momentum changes

This equation can be expressed in words:

Force is equal to rate of change of momentum.

A bobsleigh slows down from 150 km/h to 30 km/h in 15 s. If the sleigh has a total mass of 500 kg how large is the braking force?

A drag racing car of mass 600 kg accelerates from rest to 200 km/h in 5 s. What is the force being generated by the car's rocket engine?

What is the average force applied by a bow string to a 0.1 kg arrow if it accelerates to 20 m/s in 2 s? How large would the force need to be if the arrow achieved the same velocity in 1.5 s?

IMPULSE

In many sports, e.g. tennis, cricket, badminton, golf, soccer, rugby, ice-hockey, we hit or kick objects in order to change their momentum. The time during which a force is being applied, is very short, possibly as little as one thousandth of a second.

By rearranging the equation

$$F = \frac{mv - mu}{t}$$

to $Ft = mv - mu$

we can see that the size of the change in momentum depends upon the product of F and t. This quantity is known as the *impulse* of the force. If we wish the impulse of a force to be large we must pay close attention not only to the size of the force we are applying but also to the time during which it is applied.

Every golfer wants to be a big hitter 'just like Jack Nicklaus'. Consequently most golfers try to hit the ball too hard (to make F as large as possible). But they need also to bear in mind that the contact time (t) between the ball and the club is also important.

The contact time between a golf ball and a club is very short, but, as this high-speed photograph shows, the force exerted on the ball by the club has quite a marked effect.

One of the first lessons a novice golfer has to learn is not to 'hit at' the ball but to hit or sweep 'through it'. Phrases such as these are used to encourage the golfer to 'follow through' with his shot. By doing so the contact time is increased, and so the golf ball is given a larger impulse and is likely to travel further.

Most golfers try to hit
the ball too hard

The follow-through after
a golf shot

A similar situation arises in rugby. If the goal kicker follows
through, the contact time between ball and boot will be
larger and the ball will travel much further.

The follow-through after
a kick in rugby

In order to throw a shot a long way it must be given a large
impulse. The average force applied during the throw will be
determined by the strength of the athlete, but the time for
which it is applied will be decided by his throwing technique.
By beginning the actual throw from position A and releasing
the shot in position B the athlete has remained in contact
with the shot for a long period of time whilst applying a
force.

Putting the shot

A   B

31

Throwing the javelin

Why do javelin coaches insist that athletes begin the throw from this position?

Throwing the discus

Explain why the discus thrower begins to accelerate the discus from this position.

ACTION AND REACTION

Look carefully at the two ice-hockey players opposite. The player on the right is about to push the player on the left. If both players have the same mass and there is negligible friction between their skates and the ice what would you expect to see immediately after the push?

What is about to happen here?

If the player on the left had twice the mass of the player on the right how would the situation be changed?

Even without much thought most of us can see what will happen here. But can you explain why?

In order to step off his boat the rower above must move his body forward. He does this by pushing with his leg muscles. Unfortunately, as he pushes himself forwards he also pushes the boat backwards.

If we could repeat this disembarking with different rowers and boats, taking measurements of both their masses (m) and their velocities (v), we should discover that the momentum (mv) a rower gives himself as he steps off the boat is equal to the momentum given to the boat, i.e.

$$m_1 \times v_1 = m_2 \times v_2$$

Newton explained this situation and that of the two ice-hockey players with his Third Law of Motion which states that

To every action there is an equal and opposite reaction

In clay pigeon shooting, when the trigger is pulled the shot is ejected at great speed from the barrel whilst the gun itself is made to move in the opposite direction. This reaction to the ejecting of the shot is called the *recoil*.

Calculate the velocity $m_1v_1 = m_2v_2$ of the recoil, using the equation v_2:

Typical values are:

Mass of shot $(m_1) = 0.3$ kg
Velocity of shot $(v_1) = 380$ m/s
Mass of gun $(m_2) = 3$ kg

Why would a gun manufacturer want to keep the recoil of a gun as small as possible? Suggest two ways in which the designer of a gun could achieve this.

All swimmers use the principle expressed by Newton's Third Law. In order to move themselves forward they push the water behind them.

To swim quickly in a race a competitor must use a swimming technique that allows him or her to push the maximum amount of water backwards in the shortest possible time.

(a)   (b)

Breast stroke:
(a) straight arm pull
(b) bent arm pull

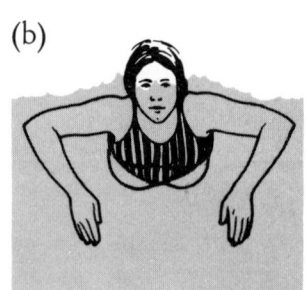

The arm action of swimmer (a) opposite is not very efficient compared with that used by a top-class swimmer. This is the arm action used by swimmers of 50 or more years of age.

As swimmer (a) pulls with her arms for much of the stroke she is pushing water sideways. Effort and time are wasted, as this does not contribute to her forward motion.

Swimmer (b) uses a slightly different arm action called the 'bent arm pull'. When her outstretched arms are slightly further apart than shoulder width she pulls backwards and slightly downwards. Most of her effort and time is now spent propelling the water backwards, the reaction to which moves her forwards.

(a)  (b)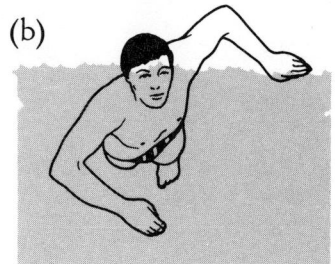

Front crawl:
(a) straight arm pull
(b) bent arm pull

Compare the two swimming strokes shown above. Then try to explain why the bent arm style propels a swimmer through the water much more quickly than the straight arm pull.

Hurdlers, when developing their technique, need to be aware of 'actions and reactions' and how to deal with them.

As the hurdler swings his trailing leg over the hurdle the upper part of his body will experience an opposite rotation. If

Clearing a hurdle

he does nothing to counteract this rotation he will land facing slightly to one side of the direction in which he is running and may overbalance. To overcome this problem he thrusts forward his left arm (right arm if the trailing leg is the right leg) with the intention of setting up an equal but opposite rotation of the upper body.

As the hurdler swings his trailing leg over the hurdle he also swings his leading leg down towards the ground. This results in a backward rotation of the upper body which could cause him to land leaning backwards and totally out of balance. To overcome this problem he leans forward 'into' the hurdle at take-off, so that the backward rotation of his upper body simply removes his forward lean. A further advantage of leaning forward at take-off is that his centre of mass is kept low.

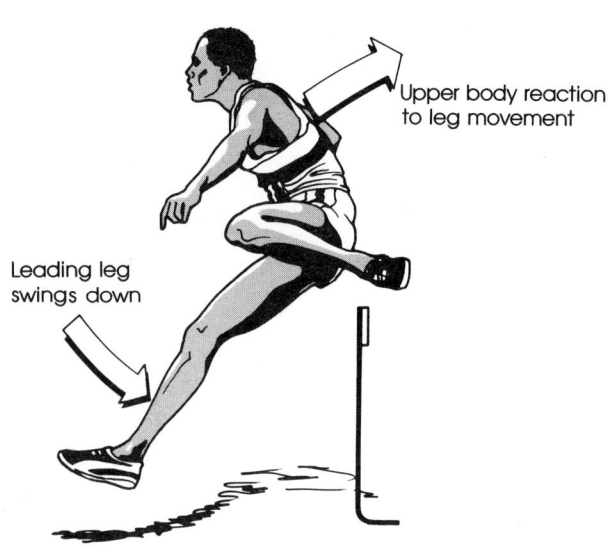

Side view of the leading leg being swung down, showing the reaction of the upper body

Upper body reaction to leg movement

Leading leg swings down

QUESTIONS ON CHAPTER 2

1 What two properties of an object does momentum depend upon?

Name two sports in which it is

(a) An advantage for an object or person to have a large momentum.

(b) A disadvantage for an object or person to have a large momentum.

2 Explain why this magician is able to remove the table cloth without breaking the plates.

3 Describe five sporting situations in which an object's momentum is changed. What, apart from the change in momentum, do all these situations have in common?

4 Explain clearly how this rower is able to move his boat through the water. What must he do if he wishes to move more quickly?

5 Look again at the first photograph on page 1. Is the racket squashing the ball or is the ball deforming the strings of the racket? Explain your answer.

6 What are 'inertia' seatbelts? Why are they so named?

7 Explain why a cyclist could be thrown over the handlebars of his bicycle if he were to apply his front wheel brake suddenly.

WORDFINDER ON CHAPTER 2

Trace this grid on to a piece of paper (or photocopy this page — teacher, please see the note at the front of the book). Then try to find as many of the words listed below as you can. They may read across, down, backwards, upwards or diagonally. Ring each word as you find it. The first has been ringed for you already.

```
E  S  L  U  P  M  I  A  R  E  C  O  I  L
O  P  P  U  L  L  S  E  D  R  P  Z  G  Y
J  E  H  M  C  M  D  R  V  P  R  U  D  L
A  E  G  I  L  R  A  K  O  E  N  Q  S  I
V  D  T  O  T  G  A  S  F  F  L  H  U  H
E  Z  R  E  M  M  I  W  S  S  O  P  C  R
L  E  A  C  B  T  L  U  L  T  N  N  I  E
I  Q  T  R  E  A  C  T  I  O  N  I  J  L
N  U  S  O  L  S  L  E  N  P  E  T  A  D
G  O  L  F  I  W  F  K  E  T  W  C  I  R
Y  X  A  D  G  N  I  C  A  R  T  A  T  U
B  M  U  T  N  E  M  O  M  I  O  G  R  H
G  J  Q  B  O  A  T  U  O  B  N  U  E  O
U  R  E  M  M  A  H  N  S  P  W  R  N  E
R  Y  T  I  C  O  L  E  V  A  E  M  I  T
```

START	INERTIA	DISCUS
STOP	MASS	ACTION
SPEED	DRAG	EQUAL
PUSH	RACING	OPPOSITE
PULL	MOMENTUM	REACTION
HIT	FORCE	BOAT
TIME	NEWTON	GUN
RUGBY	IMPULSE	RECOIL
LINE	GOLF	SWIMMER
HAMMER	SHOT	CRAWL
VELOCITY	JAVELIN	HURDLER

MOVEMENT THROUGH THE AIR

AIR RESISTANCE OR DRAG

Using air resistance to advantage

An example of the disadvantages of air resistance

The external forces experienced by objects as they travel through the air can have a considerable effect on the way these objects move. Sometimes the effect of these forces can be used to our advantage. Sometimes it is important to reduce these effects to a minimum.

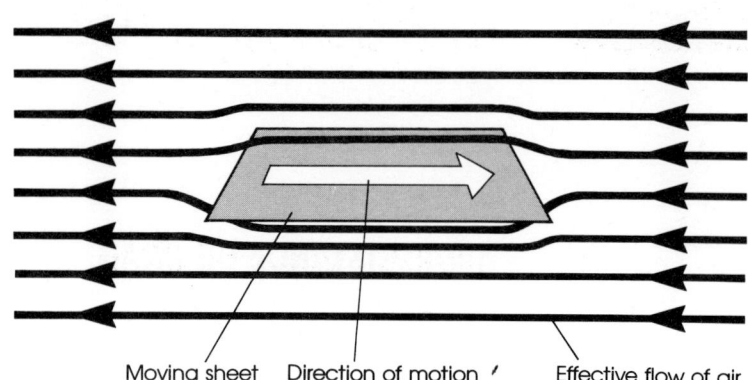

Smooth flow of air over a piece of card

Moving sheet   Direction of motion   Effective flow of air

If this large piece of card is moved as shown in the diagram, it will 'cut' through the air very easily and experience little resistance.

If, however, the sheet is tilted as shown below, it will experience considerable external forces. Below and in front of the sheet, air molecules are being pushed closer together than normal, creating a region of high pressure. Above and behind the sheet, air molecules are left more spread out and so a low pressure region is created. The result of this pressure difference is a force, part of which resists the forward motion, part of which creates 'lift'.

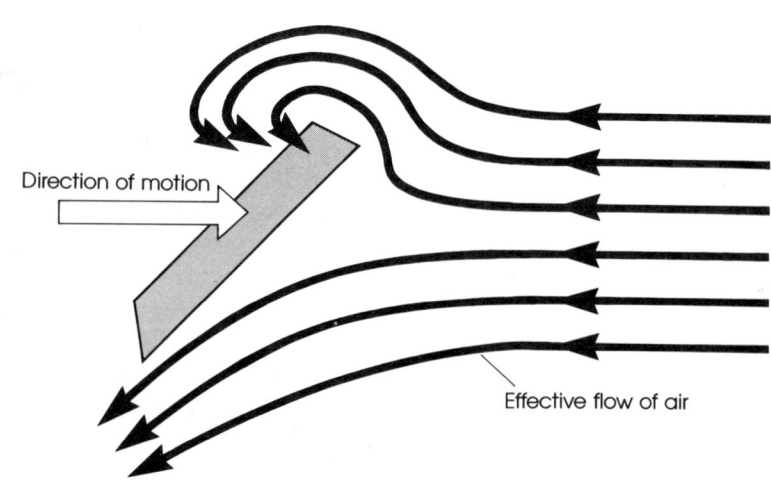

Tilting the card creates drag and lift. (See next page.)

Direction of motion

Effective flow of air

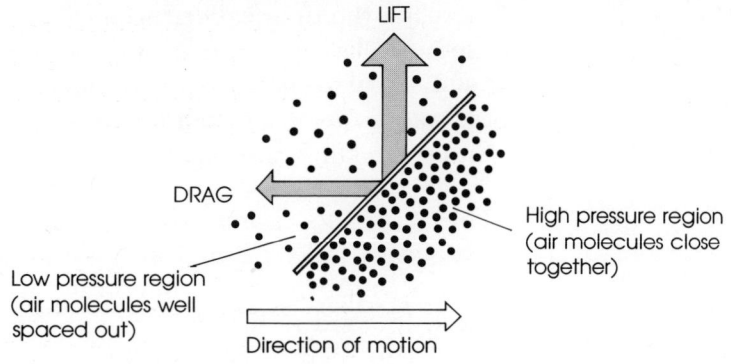

LIFT

DRAG

High pressure region
(air molecules close
together)

Low pressure region
(air molecules well
spaced out)

Direction of motion

If the sheet is moved as shown in the diagram below, large numbers of air molecules are pushed together, a large pressure difference is created and consequently the air resistance (more usually called *drag*) is much greater than in the two previous examples.

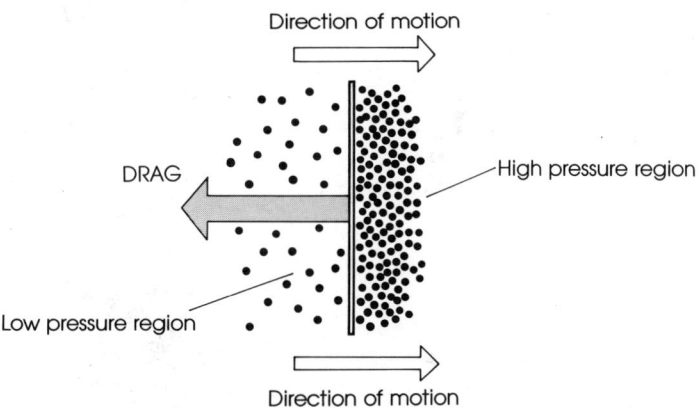

Direction of motion

DRAG

High pressure region

Low pressure region

Direction of motion

Tilting the card into this position increases drag but eliminates lift

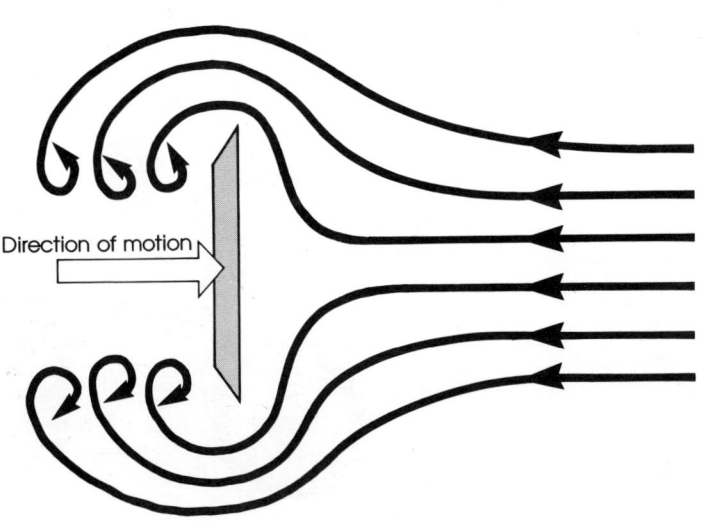

Direction of motion

Usually the drag experienced by an object increases as its speed, relative to the air, increases. If, therefore, sportsmen and sportswomen intend to move themselves or objects through the air at high speeds, the effects of these forces must be taken into account.

Motorcyclists racing round a circuit will at times be travelling at speeds in excess of 160 km/h (100 m.p.h.). Air resistance will then obviously be important. To minimise it, the riders lean forward rather than sit upright, and fibreglass fairings are fitted to their machines, so that they can cut through the air more easily. Machines such as these are said to be *streamlined*.

Because of its stream-lined shape the motor-cycle avoids creating a high-pressure region

Flow of air

Fibreglass fairings

When motorcyclists approach a corner, they can use air resistance to their advantage. By sitting in a more upright position they increase drag, and so improve their overall braking performance.

The rider's upright position increases drag which helps to slow down the bike and rider

Upright position increases drag

DRAG

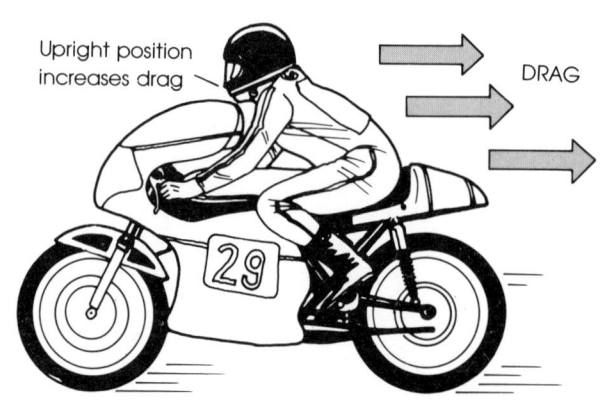

In ski-jumping, the aim of each competitor is to jump (and glide) as far as possible. In this sport, more than in most, an understanding of drag and lift, created by the air, is vital. In order to achieve maximum distance, a competitor must leave the ramp at as high a speed as possible. This means that he or she must slide down the slope in such a way that both body and skis produce little resistance.

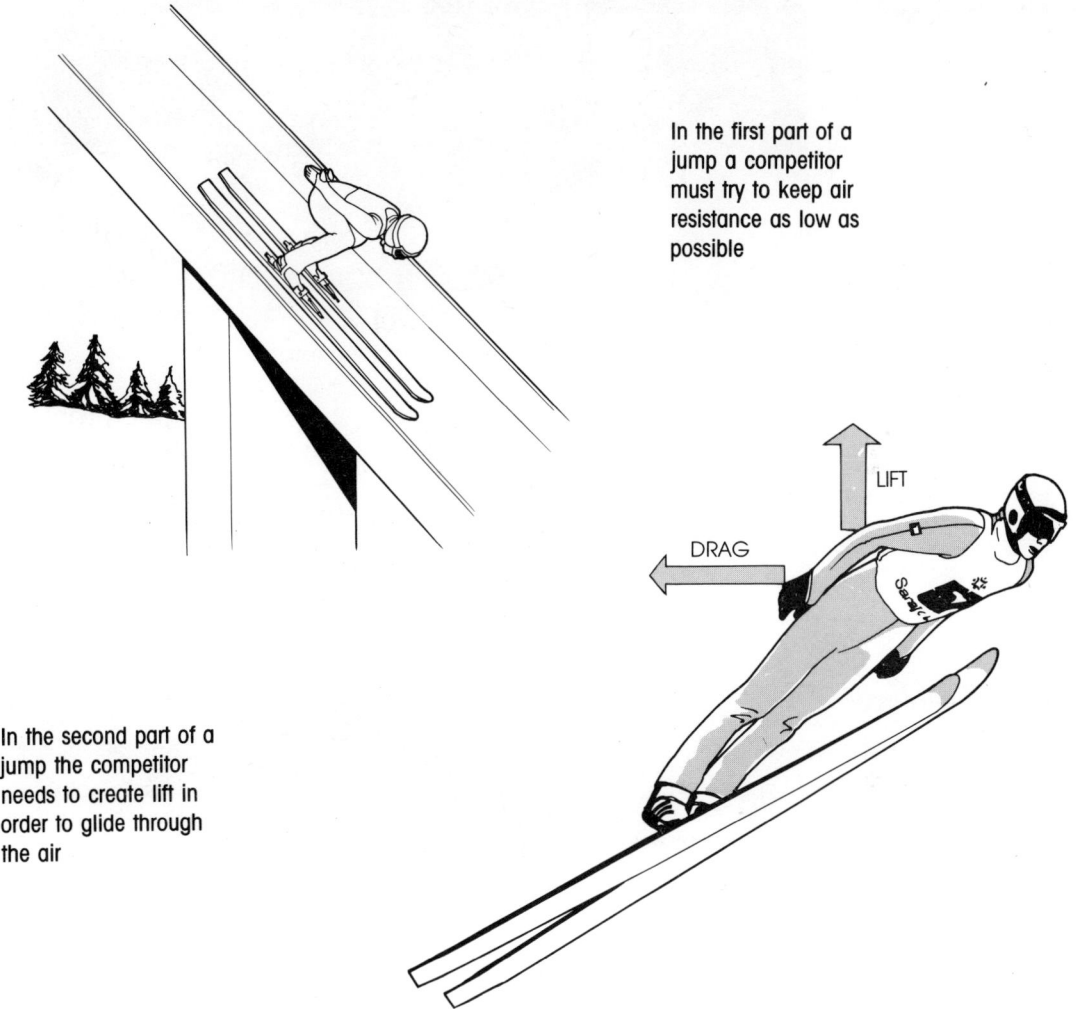

In the first part of a jump a competitor must try to keep air resistance as low as possible

In the second part of a jump the competitor needs to create lift in order to glide through the air

Once in the air, however, the situation is completely different. In order to glide as far as possible, a ski-jumper needs the surrounding flow of air to give lift. At the same time though, resistance to forward motion needs to be kept low. Thus the angle at which ski-jumpers set themselves as they glide through the air achieves a compromise between these two requirements. If the angle is incorrect, the results can be spectacular!

Explain what might happen if a competitor (a) leans too far forward or (b) does not lean forward enough.

43

AEROFOILS

A modern racing car with aerofoils

In contrast to the ski-jumpers, racing drivers face the problem of keeping the tyres of their cars firmly in contact with the ground. In fact what they require is 'inverse lift', i.e. a downward force. This can be created with *aerofoils*.

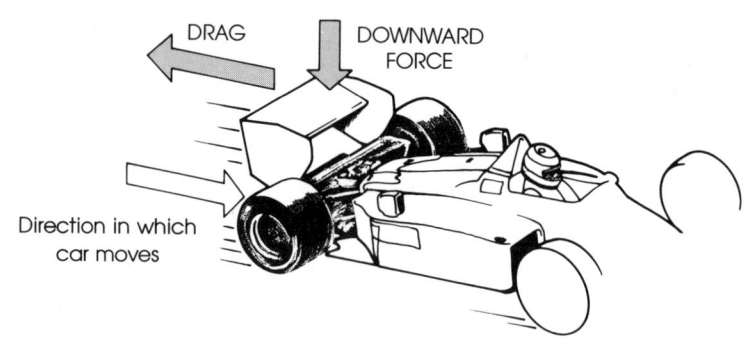

A racing car's aerofoils create a downward force, improving contact with the track surface

As the car moves through the air, a high-pressure region is created in front and above the aerofoil and a low-pressure region is created behind and below it. Provided that the aerofoils are correctly adjusted, this pressure difference will give rise to a force which helps to improve the car's grip on the road, but does not create too much drag.

Look carefully at the diagrams and photographs of various sports in this book. In which sports is air resistance (drag or lift) (a) used to advantage, (b) a definite disadvantage? How do the sportsmen and sportswomen concerned try (a) to maximise, (b) to minimise these effects? Explain each of your answers.

TURBULENT WAKES AND DRAG

If an object such as a ball moves through the air slowly, the air flows around the ball in a predictable, smooth manner. This is known as *laminar flow*.

A ball moving through the air slowly does not create turbulence

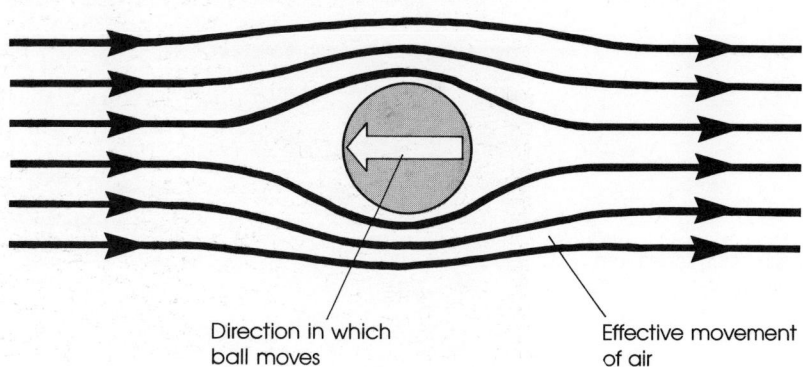

Direction in which ball moves

Effective movement of air

If the ball is moving much more quickly, an irregular air flow pattern results. This swirling movement of air behind the ball is called a *turbulent* or *eddying* wake. The ball, in creating this wake, loses energy, and therefore some of its speed. As the speed of the ball increases, the region where the wake is created moves further forward. The width of the wake, therefore, increases and the drag on the ball becomes greater.

As a ball moves quickly through the air it experiences some drag due to the turbulent wake it creates behind it

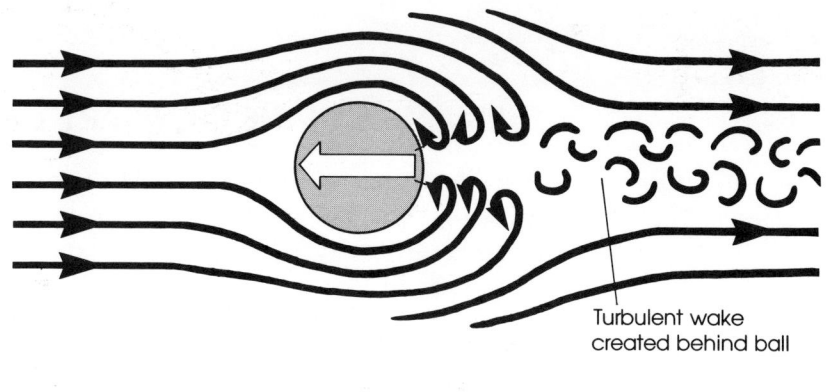

Turbulent wake created behind ball

As the speed of the ball increases the wake becomes wider and drag increases

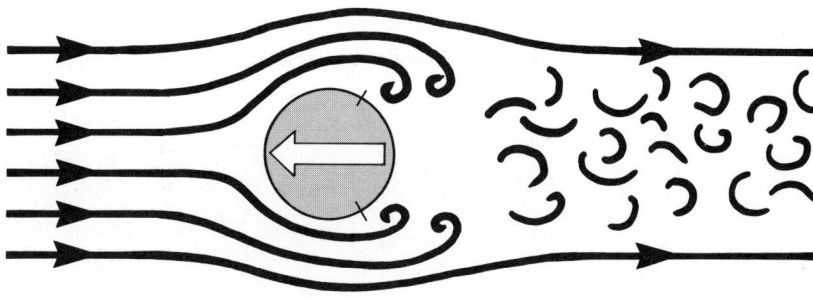

Nowadays, when a new car is being designed, a lot of attention is paid to the shape of the front and back. The front must have a shape that is streamlined, and encourages the air to flow around it easily. The rear of the car must have a shape that produces a wake that is as narrow as possible. In all new cars, the 'drag factor' is an important consideration.

A car being tested in a wind-tunnel

TURBULENCE IN THE BOUNDARY LAYER

If we look very carefully at the flow of air over the ball shown on the previous page, we shall see that a thin layer of air immediately next to the ball tends to cling to its surface. This layer of clinging air is known as the *boundary layer*.

At low speeds this clinging creates narrow wakes, whilst at higher speeds the separation of the boundary layer and the beginning of the wake occur further forward, and so a wider wake is created.

If, however, the surface of the ball is not smooth, the air in the boundary layer may become turbulent. This turbulent boundary layer tends to cling to the ball a little more than before and so the size of the wake behind the ball is reduced. Less energy is now lost in creating the wake and so the ball experiences less drag.

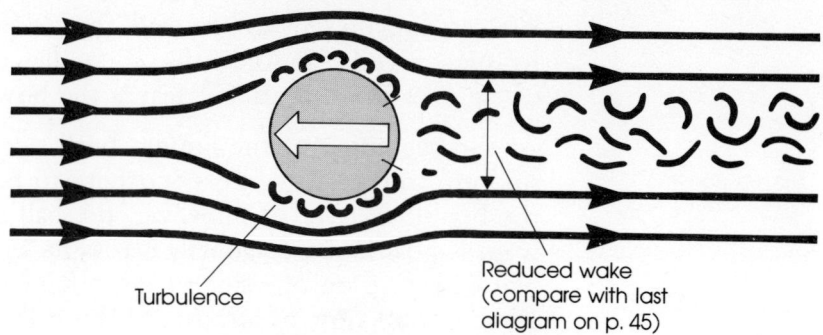

Turbulence in the boundary layer reduces the size of the wake and therefore the drag on the ball

Turbulence

Reduced wake (compare with last diagram on p. 45)

Several sports take advantage of this effect. For example, the dimpling on the surface of a golf ball causes turbulence in the boundary layer, hence a narrow wake. There is therefore less drag, and the dimpled golf ball is able to travel much further than a smooth golf ball.

Dimpling on golf balls

'SWING BOWLING'

Swinging Wood does it again!

At the Oval today, the Australian tourists found the whirlwind bowling of J.P. Wood fiercely unplayable.

This attack began early this morning, hammering home until the tourists were all out for 150 runs. Wood taking 7 wickets for just 50 runs. The remarkable medium-paced swing bowling style of Wood had

We can explain why a 'new ball' in cricket can often be made to 'swing' whilst an 'old ball' cannot by considering the flow of air over the ball after it leaves the bowler's hand.

At the beginning of the innings both sides of the ball will be smooth, and we should expect there to be no turbulence in the boundary layer. If, however, the ball is bowled so that its seam is positioned diagonally across its flight path, turbulence will be set up on one side of the boundary layer. This asymmetrical flow of air around the ball causes it to swing.

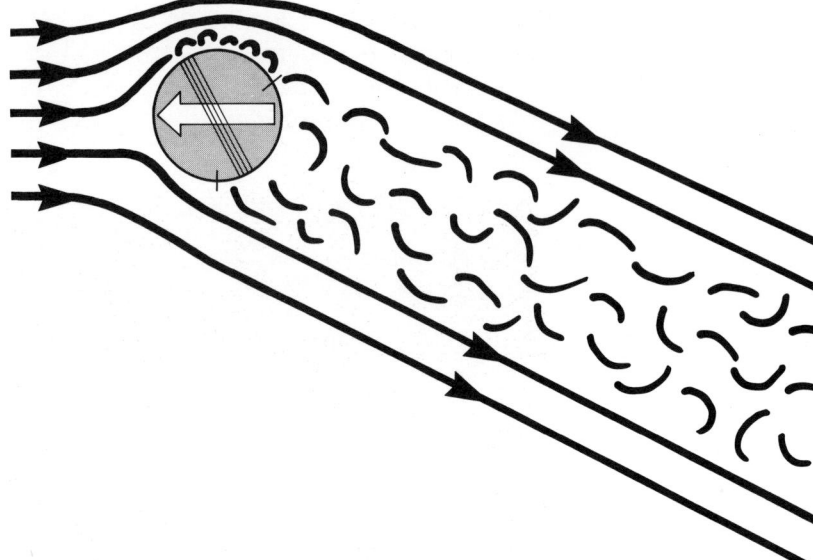

The asymmetrical flow of air over the ball causes it to swing

The forces created by this asymmetrical flow of air are quite small. Nevertheless they are sufficient to make the cricket ball move the small distance necessary to make a batsman play and miss.

Swinging the ball

After perhaps twelve or so overs or more, despite constant polishing, both halves of the ball will have become sufficiently rough to set up turbulence on both sides. The flow of air is now symmetrical, and the ball therefore stops swinging.

BERNOULLI'S PRINCIPLE

350 tonnes, and yet able to fly like a bird. How is this possible?

ACTIVITY 5

1) Take a sheet of paper and hold it close to your lower lip as shown in the diagram.

2) Blow gently over the top of the paper. What happens?

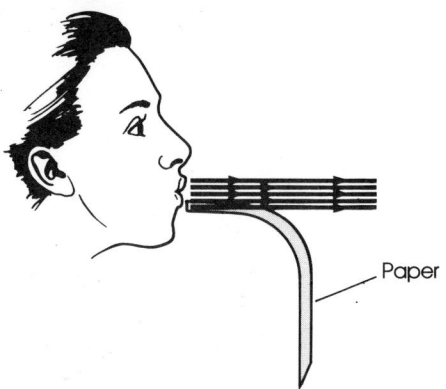

Paper

You should discover that the piece of paper rises as you blow air over it. An Italian scientist named Bernoulli was the first to explain why this happens. Bernoulli discovered that the pressure of air, when it was flowing smoothly, changed as its velocity changed. The faster the air moved (without becoming turbulent) the lower the pressure became.

Bernoulli's Principle

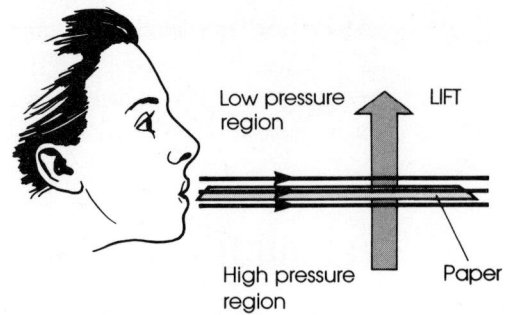

Low pressure region

LIFT

High pressure region

Paper

If we now apply this principle to our experiment, we can see that there is a pressure difference between the topside and the underside of the piece of paper. This pressure difference gives rise to a lifting force.

Bernoulli's Principle applied to the aerofoil of an aircraft

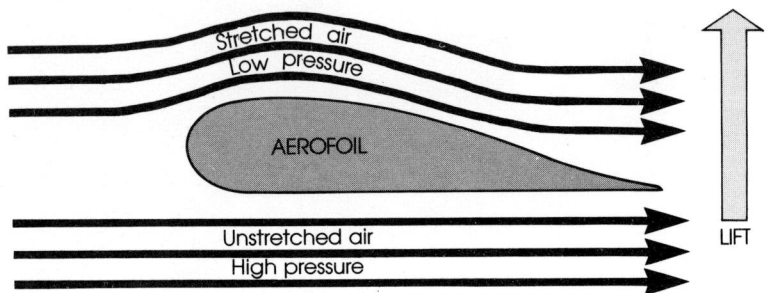

Stretched air
Low pressure

AEROFOIL

Unstretched air
High pressure

LIFT

Aeroplanes like the transport plane on page 49 rely on Bernoulli's effect to enable them to fly. The wings of all aircraft are specially shaped so that air flowing over the top of the wing is made to move faster than that passing below it. If the lift created by this pressure difference is large enough, the aircraft will fly through the air. If the lift is not large enough to support the aircraft, it will 'stall' and begin to lose height.

Suggest one reason why the aircraft might begin to lose some of its lift.

Thrown at the correct angle both javelin and discus will fly through the air

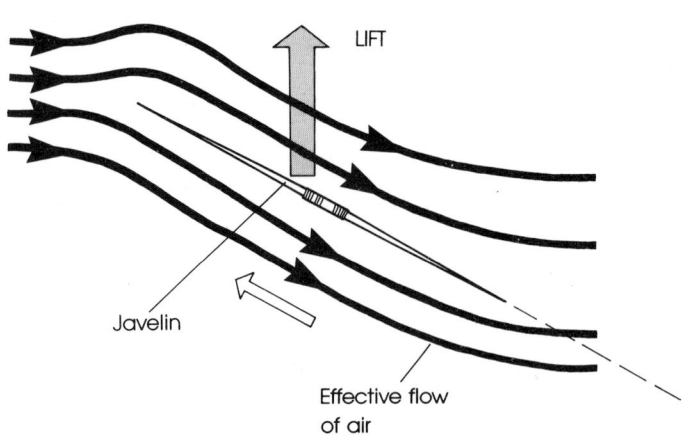

LIFT

Javelin

Effective flow of air

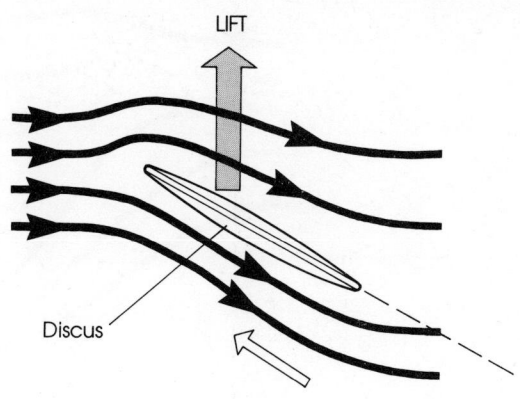

LIFT

Discus

The modern javelin and discus are both aerodynamically shaped so that they too can fly through the air. To achieve the maximum distance, it is essential that they are in the air for as long as possible. Two key factors will determine the time for which the javelin or discus is in the air:

- The velocity with which it is released (the faster the better)
- The angle at which it is released. If this is too small, there is little lift and it will not stay in the air for very long. If the angle is too large, turbulent flow is created above and behind. This disrupts the laminar flow over the upper surface, lift is lost, drag is increased, and the javelin or discus stalls. The optimum angles of release are approximately 35° above the horizontal for the javelin, and betwen 35 and 60° above for the discus.

Can you suggest one factor which might affect these optimum angles?

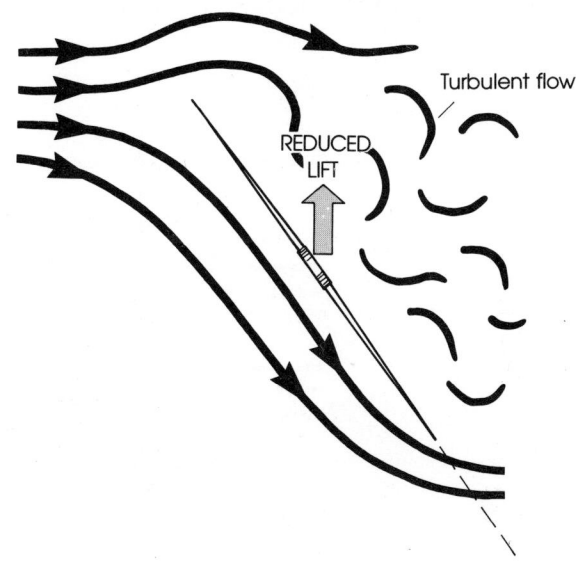

Turbulent flow

REDUCED LIFT

If thrown at too great an angle turbulent flow reduces lift and both javelin and discus (see over) will stall

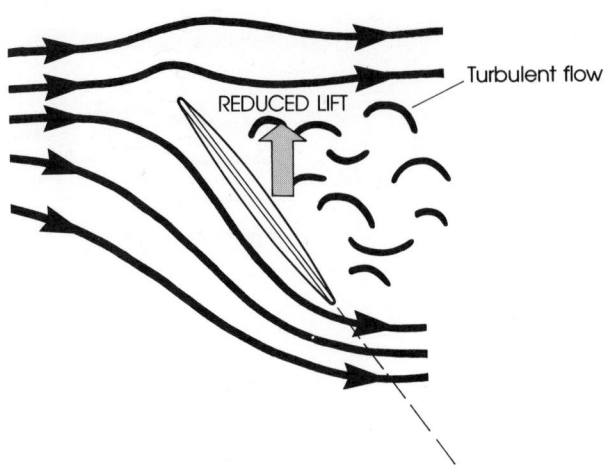

Turbulent flow

REDUCED LIFT

SPIN THROUGH THE AIR

Bernoulli's principle can also be applied to a ball travelling
through the air, but only if the ball is spinning.

Let us look at what happens to a ball as it travels through the
air with *topspin*. The air immediately next to the ball, i.e. in
the boundary layer, is being dragged around in an
anticlockwise direction. This means that at the top of the ball,
the boundary layer is moving in the opposite direction to the
main air flow over the ball. At the bottom of the ball, the
boundary layer is moving in the same direction as the main air
flow. Air will therefore flow under the ball more quickly than
it flows over it. The pressure difference this creates causes the
path of the ball to curve downwards, i.e. dip.

Side view

Topspin

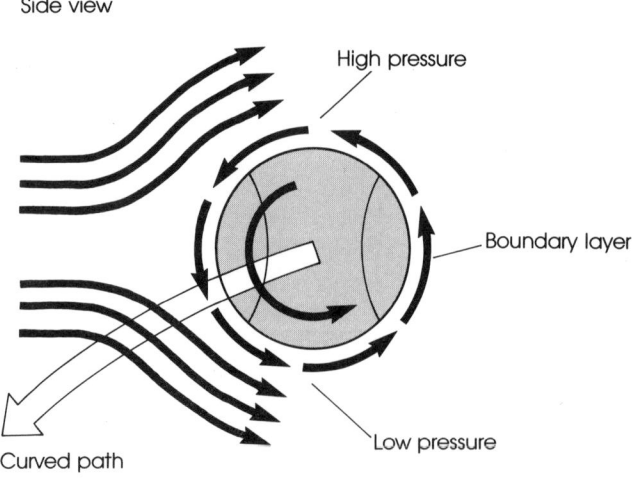

High pressure

Boundary layer

Low pressure

Curved path

If the ball is given *backspin*, the pressure difference is reversed and the ball will climb (and spend more time in the air) rather than dip.

Backspin

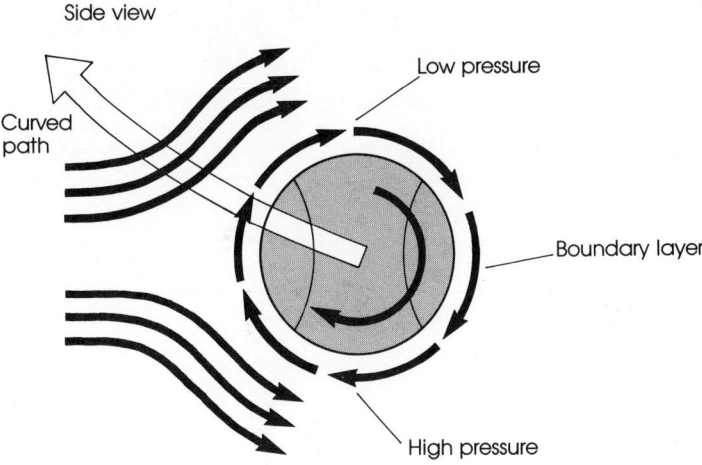

Side view

Curved path

Low pressure

Boundary layer

High pressure

Similar situations arise if the ball is made to spin in the horizontal plane.

The amount by which a ball curves in flight depends upon its rate of spin, its mass, the nature of its surface and its speed. The curvature will be greater:

- The faster the ball spins
- The smaller the mass of the ball
- The rougher and larger the surface of the ball
- The more slowly the ball moves through the air

In which part of its flight does a spinning ball tend to curve most? Why?

The table below gives some examples of the use of spin in sport.

Using Spin Through the Air

Sport	Type of curve	Action needed	Reason for use of curved flight.
Football	slice (or fade)		To pass round an opponent or deceive the goalkeeper
Tennis	dip (topspin)		Enables a player to hit the ball hard and yet keep it in court.
Golf	upwards (backspin)		To lift the ball rapidly over an obstruction such as a tree or the lip of a bunker
Golf	hook (or draw)		The steep angle of attack gives the ball backspin, causing it to rise rapidly over an obstacle, such as a tree, which is directly between the player and the green

NB The behaviour of a ball when it hits the ground often depends on the way it is spinning. This aspect of spin is discussed in the next chapter.

QUESTIONS ON CHAPTER 3

1 When an aircraft comes in to land it lowers the flaps on each of its wings. Explain in detail why it does this.

2 Manufacturers of golf balls are continually 'trying out' new patterns of dimples. Why are the dimples on a golf ball so important? What exactly are the manufacturers trying to achieve with these new patterns?

3 What is meant by 'topspin'? Describe how you would give a ball topspin? Give one example of a sport where you would use this kind of spin and explain why you would use it.

4 Name three pieces of sporting equipment that are shaped so that they move easily through the air.

5 Explain what happens when a ball is (a) sliced (b) hooked. What must happen if both of these situations are to be avoided?

WORD PUZZLE ON CHAPTER 3

The labels for this diagram can be found by taking the first letters of the answers to these questions in order.

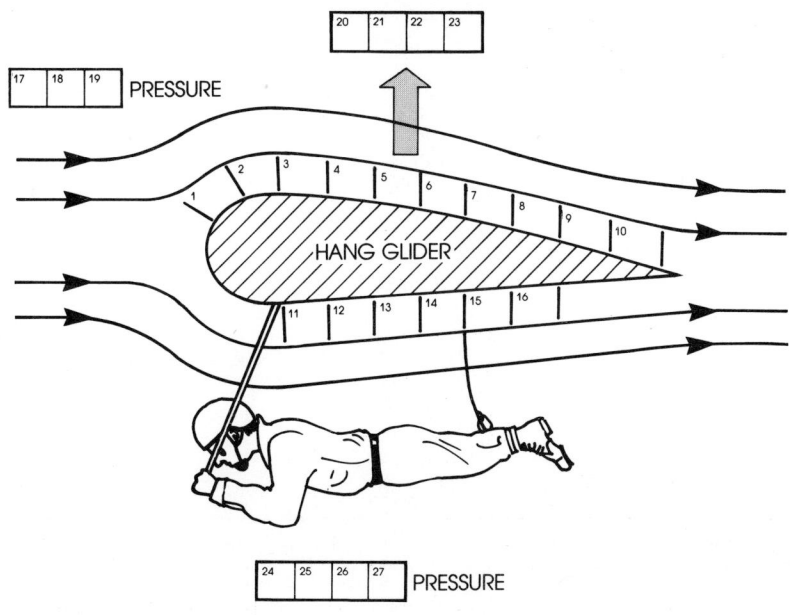

1 This kind of spin causes a ball to rise quickly.
2 A ball moving very quickly through air will create _____ currents.
3 A swing bowler wants one side of his cricket ball to be _____.
4 To travel the maximum distance, the wake behind a golf ball must be _____.
5 A ball which moves away from a batsman.
6 To put top spin on a tennis ball the racket should be moving forwards and _____.
7 A skijumper has to do this _____ in order to obtain . . .
8 _____.
9 A javelin will not fly very far if the angle at which it is thrown is _____.
10 Objects which easily cut through the air are said to be _____.
11 By creating a wake behind it, a ball is losing _____.
12 Drag is a kind of _____.
13 Racing motorcycles often have these — racing cycles don't.
14 This must lead if the sheet of paper is to cut through the air.
15 The kind of path a boomerang will follow.
16 It's the air _____ behind a racing car which creates drag.
17 Smooth flow of air.
18 The kind of cricket ball a spinner likes.
19 It's these which create the lift an aeroplane needs in order to 22.
20 Boundary _____.
21 Aerofoils _____ the downward forces on Grand Prix racing cars.
22 See 19.
23 This kind of spin will cause a ball to dip.
24 A poor golf shot that curves violently to the left (for a right-handed golfer).
25 The kind of lift created by a racing car's aerofoil.
26 How an aeroplane without engines moves through the air.
27 A new cricket ball is shiny and _____.

SPIN IN CONTACT WITH A SURFACE

ROLLING AND SKIDDING

If the rear brake is applied very gently, the rear wheel turns too slowly to match the forward motion, and skids

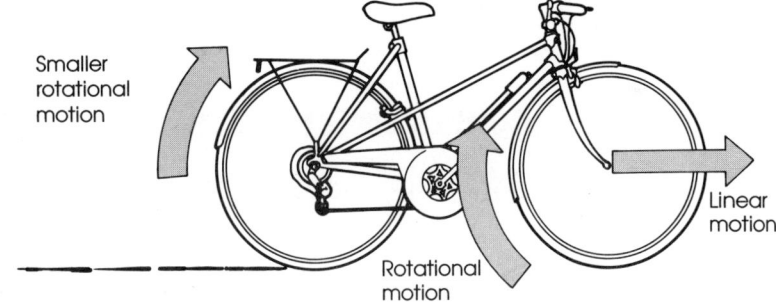

Smaller rotational motion

Linear motion

Rotational motion

If we walk along a road pushing a bicycle, the wheels of the bicycle will do two things:

- They rotate.
- They move forward in the direction in which we are pushing.

Suppose we now very gently apply the rear brake so that the rear wheel continues to rotate, but more slowly than before. We then immediately notice that the wheel begins to skid. This skidding is caused by the mismatch between the rate at which the wheel is rotating and the forward motion of the bicycle.

If a wheel tries to rotate too quickly, its rotation does not match its forward rotation and a skid results

Linear motion

Rotational motion

Larger rotational motion

A similar situation often arises at the start of a Grand Prix motor race. If a driver tries to accelerate too quickly, the wheels of the car skid until the forward motion of the car and the rate of rotation of the wheels match.

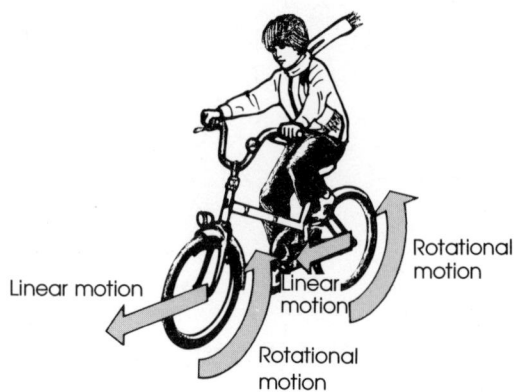

If the rate of rotation and the forward motion of the wheel match, there is no skidding

Linear motion

Linear motion

Rotational motion

Rotational motion

CONTROLLING THE CUE BALL IN SNOOKER

Just like the racing car on the previous page, the cue ball in snooker, when it starts to move, just after being struck by the cue, will often skid as well as roll across the table. By adjusting how much the ball skids and rolls a player can control its behaviour after it has hit the object ball. To understand how this is achieved, we must consider the cue ball immediately after it has been struck by the cue.

If the cue ball is struck horizontally at its centre it begins to skid across the table. Frictional forces between the ball and the table surface oppose this skidding, and because these forces are applied to the bottom of the ball, a moment is created. This causes the ball to rotate. Gradually the skidding of the ball decreases and its rate of rotation increases until there is a match between the forward motion of the ball and its rotation. The ball now moves with 'pure roll', i.e. without skidding.

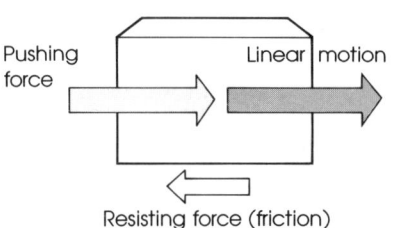

Pushing force

Linear motion

Effects of friction

Resisting force (friction)

If we push an object such as a box, it will slide or skid in the direction in which we are pushing. Friction between the bottom of the box and the ground will try to stop the box from moving.

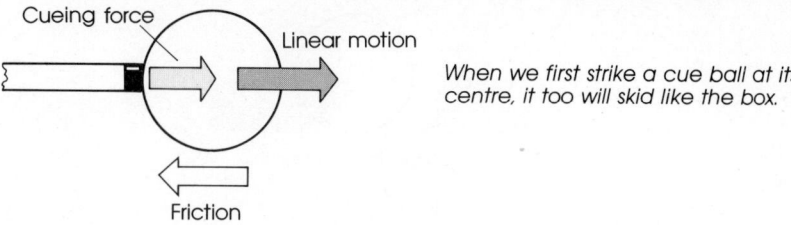

When we first strike a cue ball at its centre, it too will skid like the box.

Effects of striking the cue ball at its centre

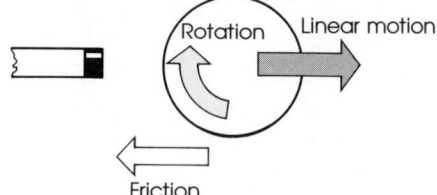

After the ball leaves the cue the only horizontal force applied to it is friction. Because this force acts at the bottom of the ball, the ball begins to rotate.

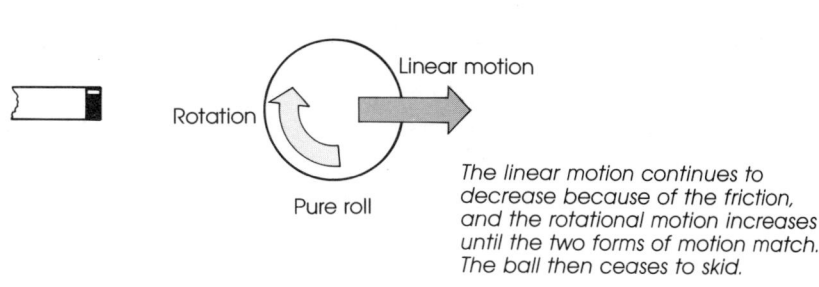

The linear motion continues to decrease because of the friction, and the rotational motion increases until the two forms of motion match. The ball then ceases to skid.

Let us now consider some of the shots used by a billiard or snooker player.

Giving the cue ball backspin (screwing back)

A snooker player often wants the cue ball to roll back towards him after it hits the object ball. To achieve this, the cue ball is struck well below its centre so that it skids towards the object ball whilst at the same time possessing a lot of backspin. When it strikes the object ball its forward momentum is lost. The frictional force between the cloth and the still spinning cue ball is now in the direction of the player, and so the ball begins to lose spin and move back along its original path. The object ball meanwhile moves away, having gained the forward momentum lost by the cue ball.

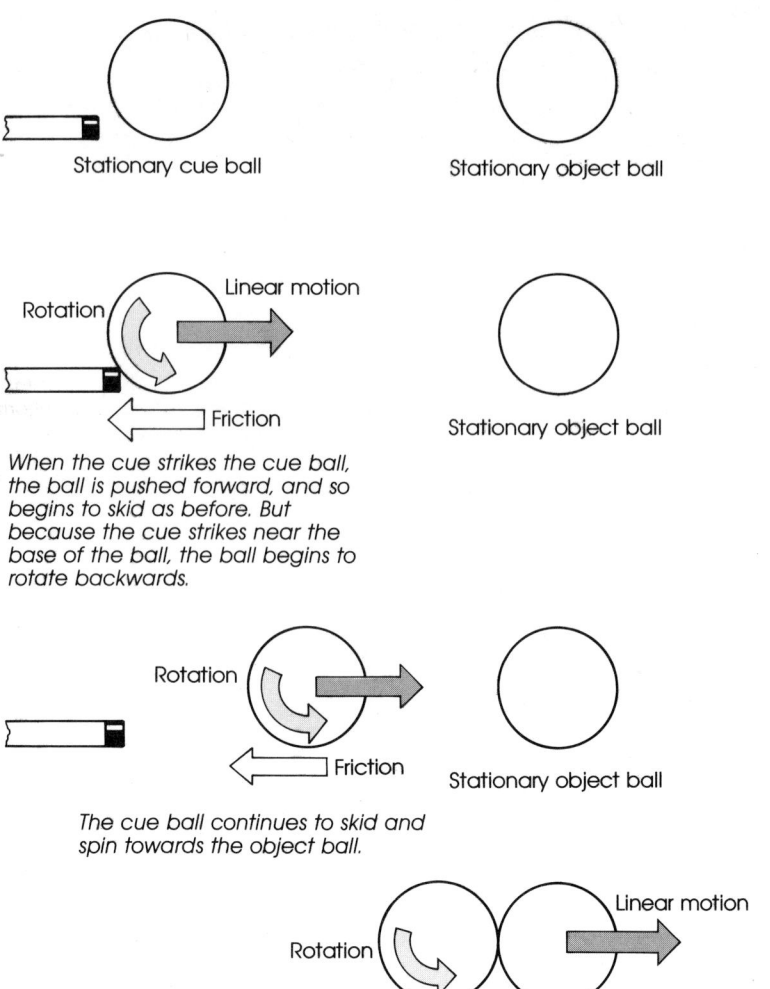

Stationary cue ball

Stationary object ball

Rotation Linear motion

Friction

Stationary object ball

When the cue strikes the cue ball,
the ball is pushed forward, and so
begins to skid as before. But
because the cue strikes near the
base of the ball, the ball begins to
rotate backwards.

The screw shot

Rotation

Friction

Stationary object ball

The cue ball continues to skid and
spin towards the object ball.

Linear motion

Rotation

Friction

When the cue ball strikes the object
ball, it stops but continues to rotate.
This spinning motion now moves the
ball back towards the cue.

Linear motion

Rotation

Rolling cue ball

Linear motion

Rotation

Rolling object ball

*The screw shot is only possible if the distance between the cue ball
and object ball is not too great. Explain why this is so.*

Giving the cue ball topspin

Sometimes a player wants the cue ball to continue to roll
away, rather than return, after it has struck the object ball.
The cue ball is then given topspin by striking the top quarter
of the ball with the cue. When the cue ball strikes the object

ball, its forward momentum is lost, and it continues to spin as before. This time, however, the frictional force between the cloth and the cue ball is acting away from the player. Therefore, as the ball begins to lose spin, it rolls after the object ball.

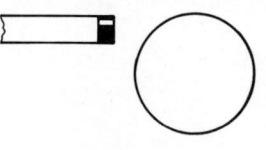

Stationary cue ball

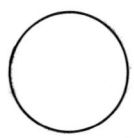

Stationary object ball

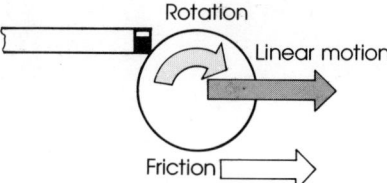

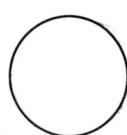

Stationary object ball

When the cue strikes the cue ball, the ball is pushed forward, and begins to skid. But because the cue strikes near the top of the ball, it gains forward spin or top spin.

Topspin

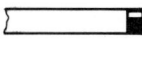

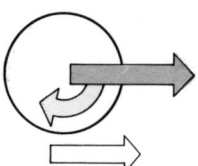

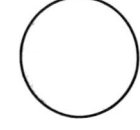

Stationary object ball

The cue ball continues to skid and spin towards the object ball.

When the cue ball strikes the object ball, the ball stops but continues to rotate. This forward spinning now moves the ball in the same direction as the object ball.

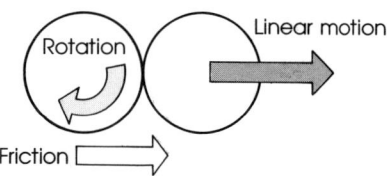

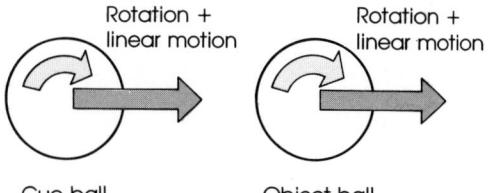

Cue ball   Object ball

All skid and no roll (the stun shot)

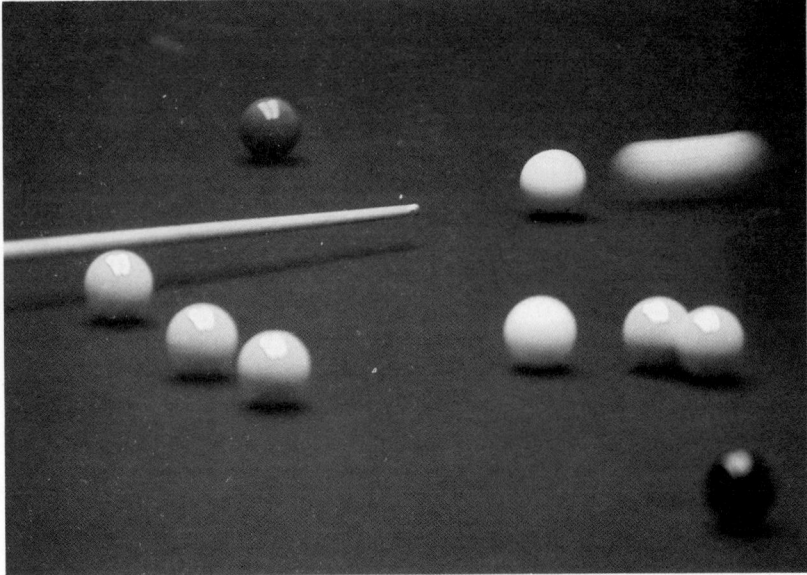

Stroboscopic photograph of stun shot

If the cue ball is struck precisely at its centre, it skids away from the player without rotating. Provided that it then strikes the object ball before the friction between ball and cloth causes rotation, all the cue ball's momentum is transferred to the object ball. Then, since there is no rotation, the cue ball stops dead. Players know this as the 'stun shot'.

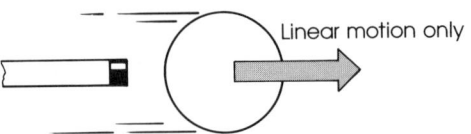
Linear motion only

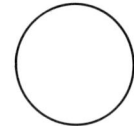
Stationary object ball

When the cue strikes the cue ball, the ball is pushed forward and begins to skid. Because the cue strikes the ball at its centre and is horizontal the ball does not spin.

The stun shot

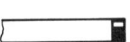

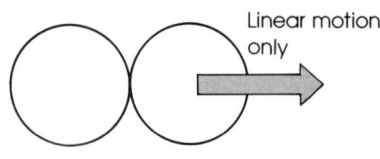
Linear motion only

When the cue ball strikes the object ball it stops. It possesses no spin and so remains on the spot where it hit the object ball.

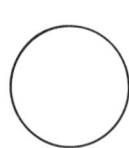

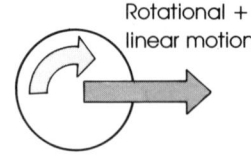
Rotational + linear motion

Stationary cue ball   Rolling object ball

All roll and no skid

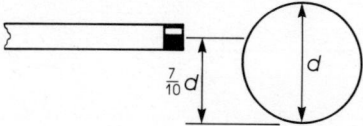

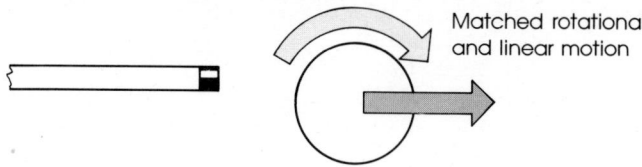

The cue ball will move without skidding if it is struck $\frac{7}{10}$ of its diameter above the table.

Matched rotational and linear motion

'Pure roll'

If a cue ball is struck $\frac{7}{10}$ of its diameter above its base it will attain a forward motion and rate of rotation which match. The cue ball will therefore not skid.

Relative heights of cushion and ball on a snooker table

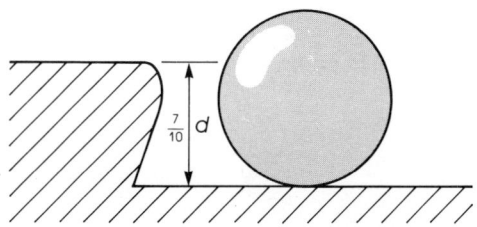

The height of the cushions around snooker and billiard tables is precisely $\frac{7}{10}$ the diameter of a snooker/billiard ball. Can you explain why this is so?

A touch of 'sidespin' (swerving the cue ball)

Until now we have only considered the effects of striking the cue ball in the vertical plane passing through the centre of the ball. Consequently the ball could only possess spin about an axis at right angles to this plane. If, however, the ball is struck to one side of this plane, the ball gains sidespin.

Suppose the cue ball is struck on the top right-hand side, with the cue angled sharply downwards. The cue ball then starts to skid away from the player whilst possessing sidespin. The frictional force between the cloth and the ball is in a direction to the right of that in which the ball is moving.

This force causes the ball to swerve to the right until it stops skidding. Afterwards the ball continues to roll in a straight line.

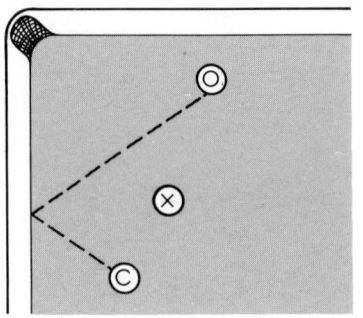

To strike the object ball (O) with the cue ball (C) a player could bounce the ball off one or two cushions, or alternatively swerve the ball around X.

Swerving the cue ball

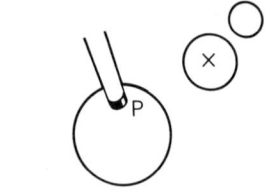

Player's view of shot

By striking the cue ball with a downward blow at point P it will be given sidespin.

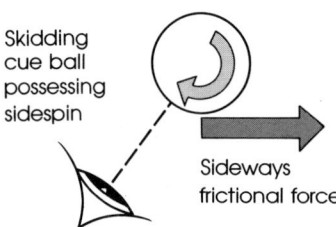

Skidding cue ball possessing sidespin

Sideways frictional force

As the ball skids forward and spins sideways it swerves (in this case to the right, a to b). When the cue ball stops skidding, it rolls in a straight line, b to c.

How would you make a cue ball swerve from right to left? Explain your answer.

BOUNCING

If we were to throw or hit a ball on to a hard flat surface, we should probably expect it to bounce off at approximately the same angle, and with more or less the same speed. In practice, though, this rarely happens. The ball, when it hits the surface, may deform, the surface itself may deform, or the ball may be spinning. All these effects influence the way in which the ball bounces. Athletes such as tennis players, bowlers and squash players are aware of this variability of bounce and use it to gain an advantage over their opponents.

Ideal bounce

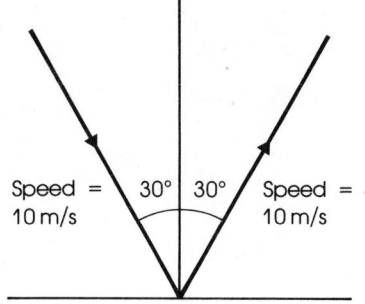

Speed = 10 m/s   30° | 30°   Speed = 10 m/s

When a ball bounces on a hard surface, it may behave like this, but often it will not, particularly if
(i) the ball is spinning,
(ii) the ball goes out of shape when it bounces,
(iii) the ground 'gives' a little when the ball bounces.

Topspin bounce

If a tennis player hits a ball so that it has a lot of topspin, he or she can hit the ball much harder than normal and still keep the ball in court. Furthermore, the behaviour of this ball, when it bounces, gives another advantage. As the ball hits the ground, it will try to skid, because its rate of rotation is too fast for its forward motion. There will therefore be a frictional force between the ball and the ground. The direction of the frictional force is the same as that of the forward motion. Consequently the ball may bounce from the ground with a larger forward velocity than expected and with a path much lower. Both effects make it more difficult for an opponent to return the ball. In simple terms, we can imagine that the ball has 'rolled through' the bounce.

Topspin bounce

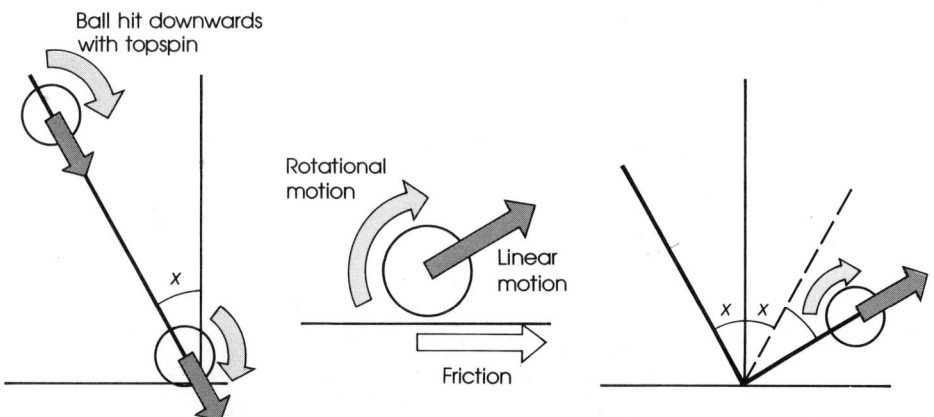

When the ball hits the ground, it is still spinning.

Friction acts on the ball because it is spinning.

This friction causes the ball to stay closer to the ground than expected.

Table tennis player
putting
topspin on ball

Backspin bounce — drop shot

If a table tennis ball is given a small forward velocity and a
large amount of backspin, the size of the frictional force
(trying to reduce the rate of rotation) may be sufficient to
cause the ball actually to bounce back over the net towards
the player of the shot. In practice, players during a game
rarely if ever encounter this extreme situation, but good
players are aware that an opponent can be beaten by hitting a
ball with a lot of backspin. Provided that the ball is not hit
too hard, much of its forward velocity will be removed
during the bounce, and the height to which the ball will
bounce will be much lower than expected.

This is an extremely useful shot to play when one's opponent
is a long way back from the table in table tennis, or near the
base line in lawn tennis. Because the ball has little bounce it is
known as the *drop shot*.

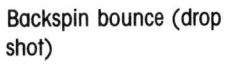

Backspin bounce (drop
shot)

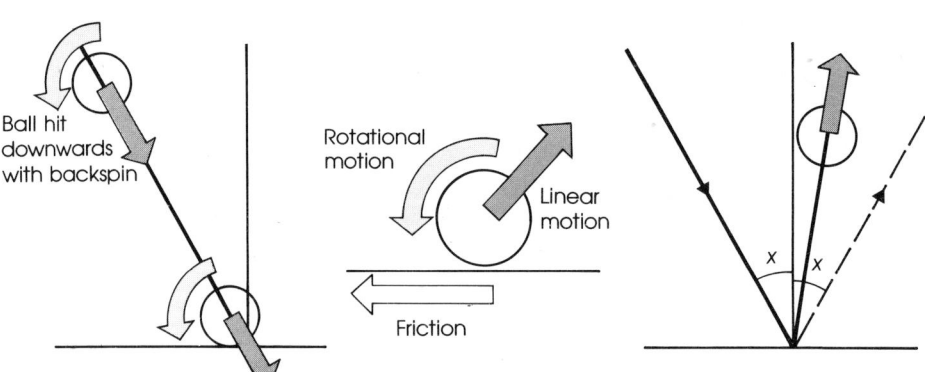

*When the ball hits the
ground, it is still spinning.*

*Friction acts on the ball
because it is spinning.*

*This friction reduces or may even
stop the forward motion of the ball.*

Variability of bounce

It is never possible to predict with 100% certainty what will happen when a ball bounces. There are many variables which affect the end result. These include:

- The nature of the ball — whether it is hard, soft, rough, smooth, heavy, light, etc.
- The nature of the surface — whether it is hard, soft, rough, smooth, etc.
- The elasticity of the ball. This depends on the materials from which it is made and how it is constructed.
- The amount of spin the ball possesses
- The direction in which the ball is spinning
- The forward velocity of, the ball

We can, however, make one generalisation. If any of these variables increases the frictional force between the ball and the ground, the ball is more likely to behave as shown in the lower diagram below.

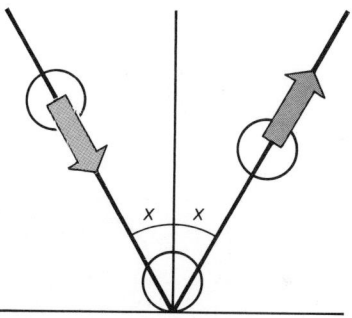

There is no friction between the ball and ground, and so the ball bounces symmetrically.

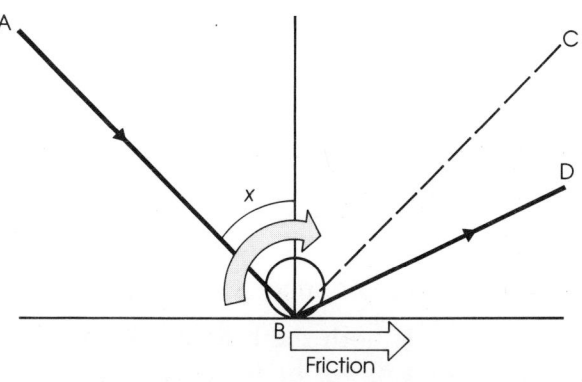

Because of friction the ball follows the path ABD rather than ABC.

Sidespin — spin bowling

If a bowler gives a cricket ball spin as he releases it, it may, when it bounces, experience large enough frictional forces to move 'off line'. Provided that the frictional force is sufficient to stop the ball from skidding, the ball then rolls sideways a little before leaving the pitch. This change in direction is precisely what is needed to deceive the batsman. The ball is described as having spun 'sharply'.

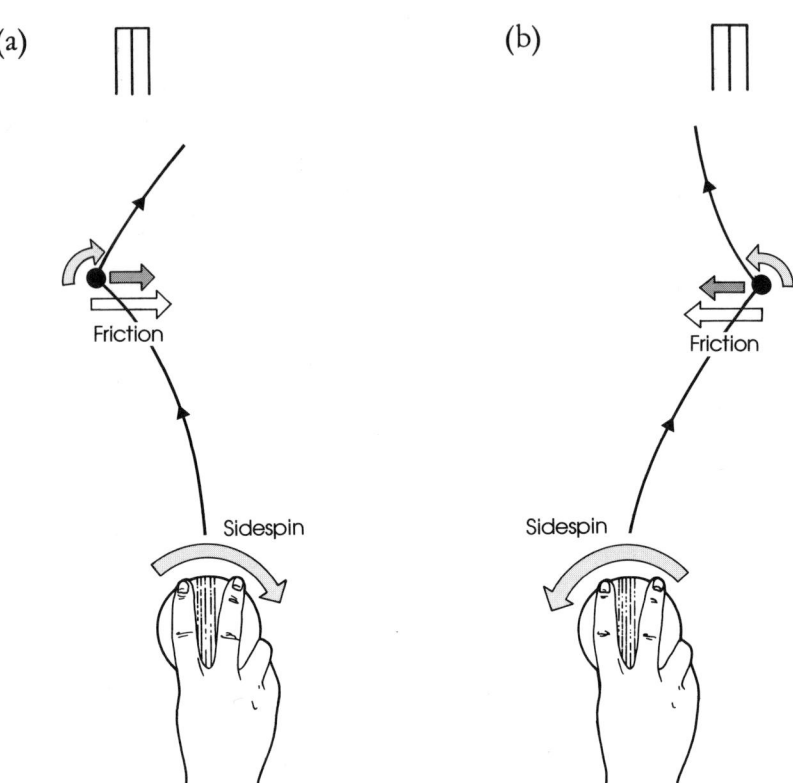

(a)

(b)

(a) Off break
(b) Leg break

Friction

Friction

Sidespin

Sidespin

The ball is given sidespin as it leaves the bowler's hand. When it bounces, the ball tries to roll to the right. This is an 'off break'.

The ball is given the opposite spin and so tries to roll to the left as it bounces. This is a 'leg break'.

Sometimes, however, a spin bowler is unable to get the ball to deviate or spin off the wicket because the ball is skidding all the time it is in contact with the ground. The effect of friction on the ball is then very small, and so the ball is seen to continue in a straight line.

If both the ball and the pitch are hard it is likely that the ball will not be in contact with the ground long enough for the skidding to be arrested. These are therefore poor conditions for spin bowling.

When the ball becomes 'old', and therefore softer, and the wicket becomes a little roughened and worn, the frictional force between the ball and ground is considerably increased. It is then easier to arrest the skidding, and so the ball may deviate by quite a large amount. These are therefore much better conditions for spin bowling.

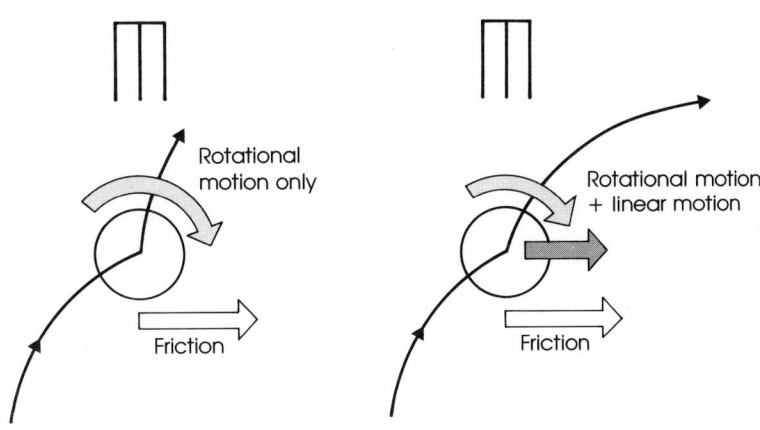

How much spin?

If the ball skids all the time it is in contact with the ground, it will not move 'off line'.

If the ball stops skidding, it will roll sideways a little, i.e. it will turn.

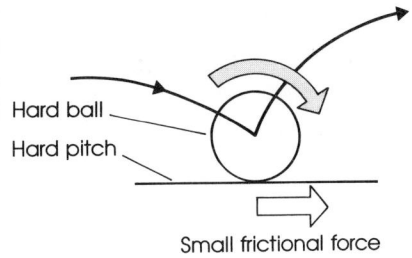

Poor conditions for spin bowling.

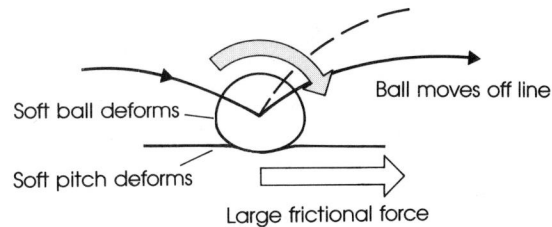

Good conditions for spin bowling.

Why is it easier to spin a 'soft' cricket ball off the pitch than a hard one?

A 'sticky wicket' is a good wicket for a spinner to bowl on. Why is the word sticky a good way to describe these conditions?

Why is 'giving the ball more air' (a higher path) likely to produce better results for a spin bowler?

Try bounce-passing a basketball as shown in the diagrams below. Throw it gently the first time and much harder the second time. Watch very carefully what happens to the ball. Then try to explain its motion.

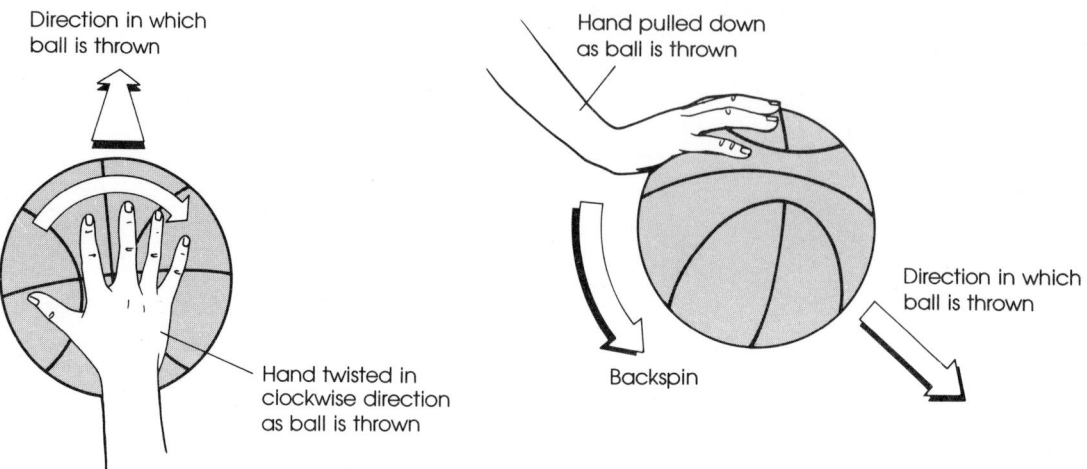

Direction in which ball is thrown

Hand pulled down as ball is thrown

Direction in which ball is thrown

Hand twisted in clockwise direction as ball is thrown

Backspin

ACTIVITY 7

Put your index finger on a spot just below the top of a table tennis ball. Press down firmly so that the ball shoots forward as it is squeezed between the table and your finger. Watch very carefully what happens to the ball. Then try to explain its motion.

QUESTIONS ON CHAPTER 4

1 If a racing driver applies his brakes too strongly he is likely to lose control of his car. Explain why this is so?

2 What kind of spin must be given to a cue ball if, after hitting an object ball, it:
(a) immediately stops?
(b) returns back along its original path?
(c) continues on away from the player?

5

FITNESS

WHAT DOES BEING FIT ACTUALLY MEAN?

It is very difficult to explain exactly what 'being fit' means. It depends upon one's age, sex, medical history and the kinds of activities in which one normally participates.

Marathon runners obviously must be fit to be able to run over 26 miles, If however, they were invited to play a sport such as squash, or participate in a weight-lifting contest, an unsuspecting spectator might judge them to be unfit. The demands made on their bodies are then totally different from those they are used to.

Fitness depends on age

Being fit means being able to cope with the demands of your sport or activity. This may mean having speed, endurance, agility or flexibility, or a combination of these.

HOW CAN I MAKE MYSELF FITTER?

There are many ways in which we could make ourselves fitter, but for most athletes getting fitter means training.

Fit for playing darts,
but what else?

TRAINING FOR ENDURANCE — STAMINA — STAYING POWER

In most sports it is an advantage for an athlete to have stamina and endurance. The principal way in which this is achieved is by exercising the heart. To understand why this improves staying power we must look carefully at how usable energy is produced inside our bodies.

In order to function properly our muscles must have a good blood supply. Dissolved in the blood are oxygen from the lungs and glucose, which is one of the end products of digesting food. These two chemicals, when they react together, produce the energy that muscles need in order to work.

If an athlete exercises groups of muscles, they need a greater supply of blood. This is achieved by the heart beating faster. A fit athlete who has trained sufficiently has a heart that has regularly experienced this increased demand for blood. It is therefore able to cope with this situation without excessive stress. By contrast the heart of an unfit athlete does find the increased demand for blood excessive and the athlete soon feels distressed.

When extra energy is demanded by the muscles of the body, the heart beats faster. When the energy demand decreases, the heartbeat gradually slows down and returns to its normal resting state. Because of his or her training, the heart of a fit athlete will return to normal much more rapidly than that of an unfit athlete. By carrying out this activity you can discover how fit you are.

1) Measure your normal (resting) pulse rate, i.e. count the number of pulses in 20 seconds and multiply this by 3 to discover the number of times your heart beats in one minute. Write this down.

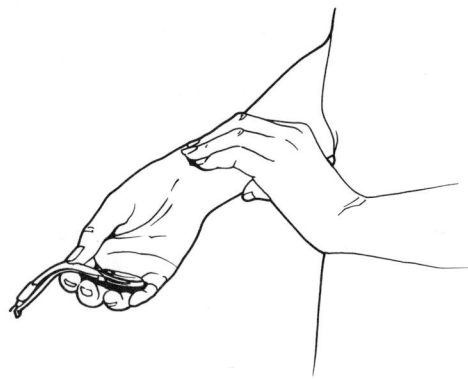

2) Run on the spot (like a sprinter) for 60 seconds.

3) Immediately you have finished the exercise measure your pulse rate.

4) Measure your pulse rate after resting for 1 minute, 2 minutes, etc.

For how long did you need to rest before your pulse rate returned to normal? Compare this time with those obtained by your friends. Who do you think is the fittest? Explain your answer.

AEROBIC AND ANAEROBIC EXERCISES

In everyday life the rate at which oxygen is demanded by the muscles of the body and the rate at which the blood supplies it are the same. Activities that can be carried out with this balance being maintained are called *aerobic* exercises.

AEROBIC ACTIVITY

If, however, the muscles of the body are exercised to such a degree that the demand for oxygen exceeds the rate at which the blood can supply it the body has to find a way of overcoming this shortfall. Activities which create this imbalance are called *anaerobic* exercises.

ANAEROBIC ACTIVITY

In events of short duration, e.g. 100 m sprint, the body allows the athlete to make greater demands on the supply of oxygen than it can cope with. But at the end of the event the athlete must repay this oxygen debt (or *oxygen deficit*), usually by breathing rapidly, i.e. panting.

In events of longer duration, e.g. 10 000 m, athletes must take great care not to establish an oxygen deficit. If they do become short of oxygen, the debt cannot be repaid, because there is no opportunity to rest. Instead, because the athlete continues to exercise, the body begins to obtain some of the energy it needs by a different mechanism. One of the by-products of this is a chemical called *lactic acid* which gradually builds up in the muscles. It is this chemical which makes our muscles feel tired and begin to ache. If athletes continue to push themselves too hard whilst experiencing these conditions their muscles may go into spasm, i.e. they can no longer control their muscles, and also experience considerable pain. This is the condition we call *cramp*. It is the body's way of stopping us from creating too large an oxygen deficit.

Excess lactic acid causes cramp. Athletes are unable to control their muscles.

Summary

- Exercising produces a stronger healthier heart.
- A strong heart copes well with the increase in blood demand which accompanies exercise.
- An oxygen deficit in the body must be paid back; otherwise there is a danger of cramp.

STRENGTH — MUSCULAR STRENGTH AND ENDURANCE

Weight-lifter

The weight-lifter in the picture above is obviously a very strong man. Most of his training routine is designed to increase the strength of the muscles in his arms, legs, chest, etc. Many athletes include some of these exercises in their training routine.

Modern athletes use the exercising equipment in a gymnasium to increase their strength. They exercise specific groups of muscles with progressively increasing weights until eventually the weight is too heavy for them to continue. By working their muscles in this way they gradually become accustomed to these heavy weights and are soon able to cope with them. This method of increasing muscle strength is based on the 'overload principle', i.e. making few repetitions of an exercise but with heavy weights. If by contrast an athlete exercises with low weights, but makes many repetitions, the exercise will improve the endurance characteristics of the muscles.

AGILITY

We describe a cat as being agile because it responds quickly to new situations and seems to be totally aware and in control of all parts of its body. Undoubtedly cats are agile from birth, and yet their agility can be seen to improve as they grow.

Cats are agile

People are not as agile as cats. They are, however, born with some inate balance, timing and co-ordination of hands and eyes. These can be developed by exercising.

Not many people would expect this 2 m (6 ft 7$\frac{1}{2}$ in) 115 kg (18 stone) man to be agile, but if he is to use his considerable strength efficiently he must move from the back of the circle to the front with great agility. His ability to do so will have come from many years of exercising and practice.

Shot putter beginning his put

Which of these athletes need to be agile if they are to succeed in their sports? Give reasons for your choice.

Boxer	*Marathon runner*
Footballer	*High-jumper*
Gymnast	*Discus thrower*
Show jumper	*Basketball player*

FLEXIBILITY

Gymnasts need to be flexible

For most athletes — even those who claim to be fit — the kind of exercise shown on the previous page is well beyond their capabilities. Nevertheless some athletes do spend a great deal of their training time developing and maintaining the natural suppleness and flexibility of their bodies. Flexibility is a measure of the range of movement of a joint. It is determined to a large extent by the bone structure and the soft tissue of the joints.

Young children are generally very flexible but as they grow older their growth slows down, their joints become a little stiffer and the range of movement becomes a little more restricted. In order to maintain or improve flexibility athletes need to follow a programme of stretching exercises.

Stretching exercises

INJURIES

Most injuries in sport are caused by sudden changes in motion. As we have seen in Chapter 1, all objects which are moving possess kinetic energy. In order to prevent injury athletes should try to avoid situations in which kinetic energy is lost or gained abruptly.

This is not a good way to stop yourself after running a 100 m race. Your kinetic energy whilst sprinting is very high, but this is abruptly lost when you hit the wall. You would almost certainly receive serious injuries.

Not a good way to stop after a race

Better, but still not good

This is still not a good way to stop yourself but you would be far less likely to injure yourself now that the wall is covered with springy padding. Once you hit the padding your kinetic energy is gradually taken from you by the springs as they become squashed. The loss in kinetic energy is much less abrupt.

When parachutists land they have the same problem as the 100 m sprinter, because they have a lot of kinetic energy to lose. Fortunately, because they are travelling vertically, they can use their legs for the same purpose as the padding above, allowing their knees and ankles to flex and absorb the shock. To reduce the effects of the impact even more, they can then roll sideways as shown below.

Parachutist landing and rolling

The sprinter below has the opposite problem to the parachutist. At present he is at rest and so has no kinetic energy. But when the gun is fired he will try to gain kinetic energy as rapidly as he can. To accelerate as quickly as possible is essential in sprinting. He will have spent many hours in the gymnasium training with weights in order to increase the strength of his leg muscles. The larger the force he can use to push himself off the blocks the greater his acceleration will be (Newton's Second Law). In order to achieve this without injuring himself he must make sure that his leg muscles are relaxed and fully warmed up by carrying out a pre-race exercise routine.

Sprinter in 'set' position

Stretching and exercising the muscles before an event reduces the chance of injury.

Why do many athletes from Britain go to other countries such as the Southern USA to train in the winter?

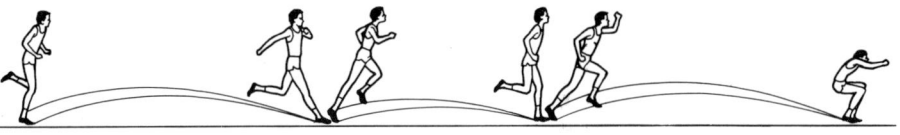

The triple jump

Look carefully at each stage of the triple jump. Can you identify four or five situations in which an athlete might be injured because of sudden changes in motion? Which is the most dangerous stage of the jump? Explain your answer.

In a sport such as cricket, which incorporates a fast-moving object, care must again be taken to avoid situations in which there is an abrupt change in kinetic energy.

This is not a good way to stop a cricket ball. It is halted far too quickly with little or no 'give'.

Not a good way to stop a cricket ball

This is a little bit better but not much. Cricket balls are heavy and can move very quickly. Consequently they have a lot of kinetic energy. In order to stop the ball without risk of injury the fielder must 'go with the ball'.

A little better, but still not good

By decelerating the ball over a longer period of time the impact remains small and is less likely to cause injury.

A much better way of stopping a cricket ball

Likewise when a boxer is caught by a good punch he must 'go with it' if he is to avoid injury.

Why are racing cars designed to crumple when they collide?

Why do boxers wear gloves?

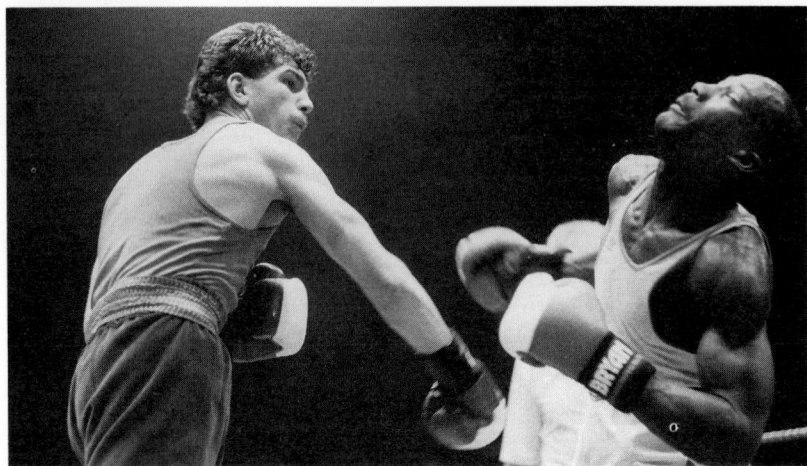

Boxer caught by a punch

Although the size of the force necessary to decelerate an object must be kept low it is also important to consider how such a force is distributed. If the force is spread over a large area there is less risk of injury. But if the force is concentrated in a small area there could be problems.

The fakir can lie on a bed of nails without injury because his weight is distributed over a large area

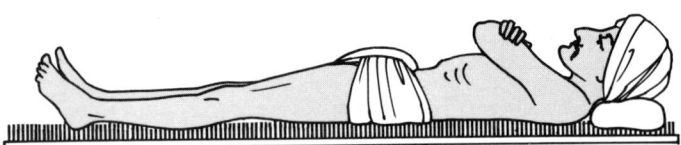

ACTIVITY 9

1) On a hard wooden or concrete floor do 10 press-ups with both hands flat on the floor.

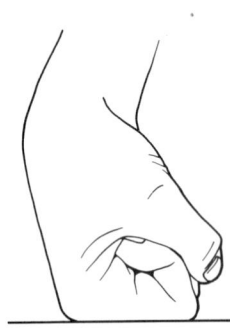

2) Now repeat the exercise with your hands made into a fist and your knuckles in contact with the floor. Try to explain any differences you feel.

When we consider the distribution and size of a force we are concerning ourselves with *pressure*.

$$\text{Pressure} = \frac{\text{Force}}{\text{Area}}$$

To avoid pain and injury we should try to spread the force over as large an area as possible. Elbows and knees are very pointed and have a small surface area. If you fall on these they will hurt and you may be injured. Try to roll just before or at impact so that the force is spread along your arms, legs and shoulders.

(a) (b)

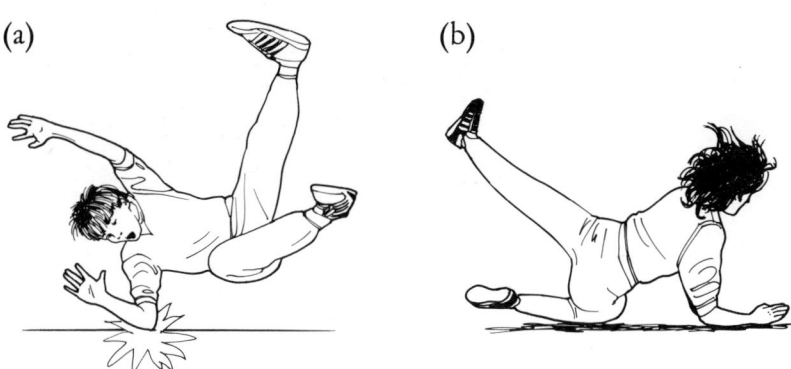

(a) Impact concentrated on a small area
(b) Impact spread over a larger area

BODY JOINTS

In order to understand how some of the most common sporting injuries occur we need to look at the way in which the joints of the body are constructed. There are two types of joint in our bodies that allow movement:

1) Hinge joints. These allow a pivoting movement in one plane. Examples of these include the elbow, the knee, etc.

Knee joint

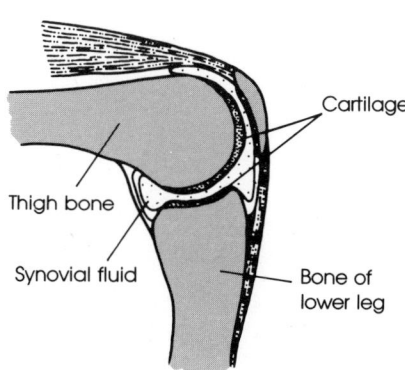

Cartilage

Thigh bone

Synovial fluid

Bone of lower leg

2) Ball and socket joints. These allow pivoting and rotational movement. Examples of these include the shoulder, the ankle, the wrist, etc.

Hip joint

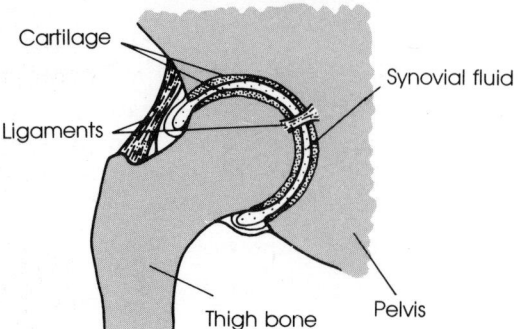

Although the joints allow different degrees of movement they do have certain common features:

- The two bones forming the joint are held in place by fibrous tissues called *ligaments*. These restrict movements that might otherwise lead to the bones moving out of position.
- The bones are connected to two muscles via a tissue called a *tendon*.
- In order to ensure that the bones can hinge and rotate smoothly there is within the joint a lubricating liquid called *synovial fluid*.
- Within the joint the bones are covered with a tough gristly material called *cartilage*. Like the synovial fluid, it encourages smooth movement between the bones of a joint.

WHAT IS A SPRAIN?

If an athlete tries to change direction quickly the knee or ankle may sometimes become sprained. A sprain occurs when an athlete tries to move the bones of a joint beyond their normal range of motion. This causes the ligaments holding the bones in place to become damaged by being overstretched.

If the movement is really severe the ligaments can actually be torn. The treatment for this injury is:

1) Initially the joint should be kept cool in order to keep swelling to a minimum.

2) The joint should be very gently exercised to encourage a good blood supply to the damaged area. NONE OF THE EXERCISES SHOULD BE LOAD-BEARING.

WHAT IS A DISLOCATION?

In some sports, particularly those in which an athlete is likely to receive a sudden jolt or knock, it is possible for one of the bones of a joint to become completely dislodged from its correct position. This kind of injury, in which the ligaments have been unable to hold the bones in place, is called a *dislocation*.

Bones which have become dislocated obviously need to be treated by an experienced physiotherapist or a doctor. Attempts by enthusiastic amateurs to correct the problem can result in the joint being more severely damaged.

PULLS, STRAINS AND TEARS

As we have already seen in Chapter 1 movement is the result of the contracting and relaxing of groups of muscles. If, however, a muscle contracts violently, either it or the tendons attaching it to the bones may be damaged or strained. Indeed, if the contraction is sufficiently large and abrupt the tissues may actually tear. If a muscle tears, there will be internal bleeding and the joint will begin to swell and bruise. This swelling should be kept to a minimum by wrapping the joint in a pressure bandage and keeping the muscle cool. Very gentle massage will encourage the flow of blood to the damaged area and will accelerate the repair process. It is not necessary or desirable to rest the joint completely, but as with sprains, load-bearing exercises should be avoided to start with.

If the tendon is torn the treatment is less straightforward. Tendons that are badly torn or ruptured will almost certainly need the movement of the joint to be severely restricted, possibly by a plaster cast. Less serious damage can be repaired by heat treatment and massage.

CARTILAGE DAMAGE

We often hear of athletes, particularly footballers, who have 'cartilage trouble' in joints such as the knee. Some of the cartilage covering the end of the bone in the knee joint can be torn and loosened by the violent twisting movements the leg is subjected to throughout a game. If the cartilage is damaged it inhibits motion and causes considerable pain and discomfort. The cartilage receives no blood supply and therefore cannot be repaired by the body. If an athlete is in pain or has the movement of a joint restricted because of damaged cartilage, the cartilage will probably need to be removed by surgery.

WHAT IS CRAMP?

Cramp is the involuntary tightening of a muscle. It tends to affect those who take part in sports which go on for a long time, e.g. marathon runners, tennis players, and long-distance swimmers. It is almost certainly caused by the presence of excess lactic acid in the muscle (see page 75), and appears to occur more frequently if the weather is either cold enough to reduce the blood circulation or so warm that an athlete loses a lot of salt and other body chemicals by perspiring heavily.

A well prepared warm-up routine will reduce the likelihood of cramp in cold conditions, and taking salt tablets before an event will reduce the danger of cramp in very warm conditions. If a muscle does go into spasm it can be treated by gently stretching it.

WHAT IS STITCH?

Although most people have experienced stitch at some time in their lives, no one knows precisely what it is. It tends to occur in unfit people who try to exercise too vigorously, too quickly, and in people who exercise too soon after eating a meal. This suggests that the pain is related to the digestive system but we do not know for certain.

To avoid stitch do not exercise too quickly, and be sure to allow your last meal plenty of time to digest (approximately $1\frac{1}{2}$–2 hours).

DRUGS IN SPORT

The use of drugs in sport today is a highly controversial issue. They can, in the hands of a qualified doctor, be extremely beneficial in the treatment of a variety of injuries and illnesses. If, however, they are misused by athletes or their coaches, the consequences can be devastatingly tragic. To understand why this is, it is necessary to understand how different drugs affect the human body.

Painkillers or analgesics

Everyone has at some time suffered from a headache. For many people the solution is to take aspirin or paracetamol. After a little while the pain subsides and life can go on as normal. Aspirin and paracetamol are painkillers. They belong to a family of drugs called *analgesics*. These affect the nervous system, blocking out the messages of pain being sent to the brain.

At first the idea of using these drugs to relieve the pain an athlete feels during training or competition seems very attractive, but it is potentially very dangerous. Athletes who are injured ought to be resting, but they may be determined to enter an event, even though they know their injury will cause pain because it has not fully recovered. They may then decide to take an analgesic to help overcome the pain. What they have not appreciated, however, is that pain is one of the body's ways of telling us that something is wrong, to stop exercising and to look after the injury. If these messages have been blocked off from the brain by the drug these athletes may without realising it aggravate the injury, possibly causing permanent damage. Therefore painkillers should not be taken for sport injuries without consulting a doctor.

Stimulants or amphetamines

In many sports stamina and alertness are two of the key qualities desired by athletes. Sometimes, however, the training schedule, which is designed to improve these qualities, can leave one tired and 'over the top'. To help overcome these feelings some athletes fall into the habit of taking drugs called *amphetamines*, often nicknamed 'uppers' because they make one feel 'high', alert and able to shrug off tiredness and fatigue.

The amphetamines do this by stimulating the heart, making it beat more quickly, and as a consequence increasing the blood pressure. These drugs, like the analgesics, affect the message reaching the brain, but causing it to work even harder. As a result the athlete could collapse from exhaustion and even die. Another danger is that these drugs are very addictive, and without them an addicted athlete soon feels depressed and weary.

Never take stimulants to improve your sporting performances. In the long run they could slow you down — permanently!

Tranquillisers or sedatives

In sports such as shooting, archery, etc. it is an advantage to be relaxed and not overexcited. Feelings of tranquillity can be induced by taking drugs called *tranquillisers or sedatives*, but once again there are dangerous side-effects. These drugs are addictive, and can cause hallucination. They can also cause drowsiness and affect reaction time and co-ordination. It is far better to have a peaceful night's sleep before an event, and awake fully relaxed, than to attempt the event while drugged with tranquillisers, and possibly sleep for a very long time afterwards.

Anabolic steroids or body builders

There is within our bodies a chemical called testosterone. It promotes and encourages protein building, i.e. body building. If athletes take a similar but artificial growth stimulant, called an *anabolic steroid*, they can use it to gain weight very rapidly. They will experience dramatic growth accompanied by substantial increases in strength. This increase in strength for an athlete who refuses to take the drug can only be achieved after many years of hard training. Using steroids to increase muscle power is very attractive to competitors in events such as putting the shot, hammer-throwing, discus-throwing, etc. But once again the use of these drugs can have tragic consequences. They can cause sterility in an athlete and there is also a serious risk of the liver being damaged.

Summary

As this section shows there are some advantages in using drugs (or 'dope') but these are insignificant compared with the dangers. Serious injury or death are the possible consequences of taking drugs. So the message is clear:

Don't be a dope.

THERE IS NO PLACE IN SPORT FOR DOPES.

QUESTIONS ON CHAPTER 5

1 Jogging is a very popular way of getting fit.
  (a) Explain what 'being fit' means.
  (b) Explain why jogging makes us fitter.
  (c) Give three everyday examples of when it is to our advantage to be fit.
  (d) Suggest three or four things we should not do if we wish to remain fit. Explain why and how they would effect our fitness.

2 For each of these different aspects of fitness give examples of two sports or activities in which they are very important.
  (a) Strength
  (b) Agility
  (c) Stamina
  (d) Flexibility

3 Write down five different types of injury athletes may experience. Explain precisely what has happened to their bodies and how they should be treated.

4 Why is it important that all athletes go through a 'warming up' routine before an event or game?

Choose one event or game and suggest what the routine might be and why.

5 Write down the names of three groups of drugs.

QUIZ ON CHAPTER 5

Answer 'true' or 'false' to the following questions. Award yourself one point for each correct answer, or see how many questions you can answer correctly in 2 minutes.

1 The fibrous tissues connecting two bones are called muscles.
2 The lubricating liquid in a knee joint is called aqueous humour.
3 The gristly tough material covering the bones in the knee is called the cartilage.
4 Sprain can be caused by athletes overloading joints.
5 It is important to keep the swelling of a sprain to a minimum. This can be done by bandaging the joint tightly.
6 Sprains should be exercised rigorously to prevent them from 'seizing up'.
7 Ligaments require a blood supply.
8 Damaged cartilage will repair itself if the joint is rested.
9 Cramp always occurs in the legs.
10 Being too hot or too cold may result in an athlete having cramp.
11 An analgesic is a pain-killing drug.
12 Amphetamines decrease the blood pressure.
13 Amphetamines are addictive.
14 Tranquillisers are not addictive.
15 Barbiturates are a kind of anabolic steroid.
16 Anabolic steroids have been used by athletes in an attempt to keep their weight down.
17 Taking a drug just once to help you perform in your sport is OK.
18 Professional medical advice should be sought before taking any drug.
19 Drugs have killed fitter and healthier athletes than you.
20 A ball and sprocket is a kind of joint.
21 To avoid injuries you should avoid concentrating forces in one area of the body.
22 Stretching exercises before an event reduce the chance of injury.
23 The bigger the force the bigger the acceleration – this is Newton's First Law.
24 Moving your hands towards a ball as you catch it will reduce the chance of injury.

25 High repetition exercises will increase muscle endurance.
26 The rapid loss of kinetic energy rarely causes injury in sport.
27 Marathon runners, weight lifters and gymnasts are all likely to be fit but in slightly different ways.
28 Acetic acid causes tiredness and cramp in muscles.
29 Anaerobic exercises use up oxygen more quickly than the blood can supply it.
30 A 100 m sprinter is performing an aerobic exercise.
31 An athlete with a high pulse rate is fitter than an athlete with a slow pulse rate.
32 A fitter heart means a fitter athlete.
33 Blood carries oxygenated food to all parts of our bodies.
34 If an athlete's lungs are small he or she usually has to pant at the end of a race.
35 None of these athletes needs agility: hammer thrower, discus thrower, boxer.

INDEX